Storm Surge

Also by William Sargent

SHALLOW WATERS
A Year on Cape Cod's Pleasant Bay

THE YEAR OF THE CRAB
Marine Animals in Modern Medicine

UNDERSEA LIFE IN AMERICA

NIGHT REEF

A YEAR IN THE NOTCH
Exploring the Natural History of the White Mountains

CRAB WARS
A Tale of Horseshoe Crabs, Bioterrorism, and Human Health

STORM SURGE

A Coastal Village Battles the Rising Atlantic

William Sargent

University Press of New England

HANOVER AND LONDON

Published by University Press of New England
One Court Street, Lebanon, NH 03766
www.upne.com

© 1995 by Francis W. Sargent, Jr.
Afterword © 2004 by Francis W. Sargent, Jr.

Maps by Kristina Lindborg Sargent and John E. Cook

Originally published in 1995 by Parnassus Imprints, Inc., Hyannis, MA 02601

First University Press of New England edition 2004

Library of Congress Control Number: 2004108857
ISBN 1-58465-406-6

Printed in the United States of America   5 4 3 2 1

The Library of Congress has cataloged the original edition as follows:

LIBRARY OF CONGRESS CATALOGING-IN-PUBLICATION DATA
Sargent, William, 1946–
    Storm surge: a coastal village battles the rising Atlantic /
    William Sargent. — 1st ed.
        p.   cm.
    Includes index.
    ISBN 0–940160–60–9 (pbk.)
    1. Coast changes—Massachusetts—Chatham (Town).
    2. Storm surges—Massachusetts—Chatham (Town). I. Title.
    GB459.4.S27 1995
    974.4'92—dc20   95–6372

Dedicated to my father

Contents

Acknowledgments

I have never been able to write a book that proceeded in a leisurely, straightforward manner. For some reason I select some small and obscure topic like a bay, or an animal, or an inlet and proceed to pick it apart to see if I can draw some broader implications. It's a method that is encouraged and works particularly well in science, but it drives most publishers to distraction. "How can I sell a book about an inlet?" . . . "How can I sell a book about horseshoe crabs," the less perceptive will declare.

Obviously it takes editors, publishers and agents of particular perception and daring to believe in such books. I have been fortunate enough to find those people. My agent, Ann Rittenberg, encouraged and helped shape this book as it evolved from a straightforward science book to a book about people. She peddled it patiently from publisher to publisher to find just the right editor. I thank her for all her efforts, and admire her for still having time to give birth to twins during the process.

An author usually doesn't thank editors who do not publish his books; however, I would like to make an exception. Both Howard Boyer of Shearwater Books and Jack McCrae of Henry Holt Company gave me encouragement when I needed it most.

Wally Exman of Parnassus Imprints has been a pleasure to work with. His light pencil, strong enthusiasm and self-deprecatory comments in the margins subtly indicated where I had made an unpardonable gaffe or written a particularly abstruse passage.

All the people mentioned in the bulk of this book have given of their time and help in recording their stories and points of view. Specifically I would like to thank Graham Giese, Dave Aubrey, Bob Oldale, Orrin Pilkey, Bert Nelson, Steve Rolfe, the Edsons, Nick and Cecilie Brown, Chris and Su-Ann Armstrong, Shareen Eldredge, Barry Eldredge, Jim Lindstrom, John Whelan, Stuart Smith, Stuart Moore, Stuart Crosby, Richard Hiscock, Dick Miller, Nick Soutter, Spencer Kinnard, Tim Wood, Tom Marshall, Jane Vollers and Paul Marinacchio for their time, help, encouragement and patience. If I have left anyone out, appeared to pass judgment, or been critical of anyone in this book, I hope you will understand it as gentle criticism done in the service of telling a good story while examining the complexities of dealing with people's lives in the face of environmental change.

Finally, I would like to thank Ben and Chappell for putting up with my endless hours of rewriting and rethinking this book. Especially I would like to thank Kristina for her love and support and for putting her own career on hold to help me write my books over these many years. I love you all.

Preface

This book is dedicated to my father. Some people may remember him as the Governor of Massachusetts, others as an early environmentalist fighting for the Cape Cod National Seashore. I remember him as a fisherman, a father, and a friend; but most of all, I remember him as a teller of stories.

You see, he loves to tell stories. Not jokes but stories. They are not grand but simple. The punchlines are often incidental or nonexistent. He is seldom the hero of his stories, and often the goat. In fact, those were the stories he loved the most, and in the retelling his goatishness loomed larger and larger.

The stories usually revolve around something amusing that had happened at the State House in Boston. In fact, I think half of the reason he liked working in the State House was that it gave him an endless repertoire of good stories.

We usually heard these stories around the dinner table. I suppose my father sometimes talked about issues or ideology or meeting someone famous. If he did I don't remember it very well. What I do remember were the stories; the stories that showed people with all their wonderful quirky humanity, stories that reflected a love of people as they are, not as they should be.

Putting these stories on paper will not do them justice. They are meant to be told, not read. The audience for these stories often knew every part of them by heart—the premise, the lines, the situation and the outcome. That was okay; it was not really the story we loved, it was the telling, the acting, the fun. It was my father taking so much pleasure from telling the story that his own laughter often made him muff the punchline. I might not do the stories justice but perhaps it will just make me feel good to remember them once again . . .

I remember my father telling of meeting an inmate at the Concord State Prison. "He was doing a little time for a few bank jobs. I liked him right away. He was as political as Patty's pig and had a great sense of humor." My father always believed that someone with a good sense of humor couldn't be all bad.

"On the way out I told the warden how impressed I was with the guy and asked when he was due to be released. 'It's funny you should ask, Governor. We've been thinking of suggesting him for early release.' Well, he did get out, and I helped get him a good job with the Department of Corrections. He went on to become an exemplary husband, father and community leader.

"A few years later I happened to be standing in line at one of those big banks in Boston. Suddenly I heard this familiar voice behind me, 'You know, Governor, in the good old days I'd just go right up to the front of the line and help myself.' Of course it was the reformed convict."

Another time my father told us about Billy Bulger, the Democratic President of the Massachusetts Senate. As leaders of opposing parties, I suppose they should have been sworn enemies but they were really good friends. "Bulger was just the Senator form Southie then, but he was bright and ambitious and funny as hell. So I was surprised when he came into my office one day to see if he could get a judgeship.

"Frankly, I tried to talk him out of it, thinking he would miss the action in the Senate. I asked him if he thought he could raise his nine kids on a judge's salary. He said he had thought about all those things but he really wanted to be a judge. 'However, Gover-

nor, I have one problem. You know my brother has a record.'
Look, I said, you're the guy I'll be appointing, not your brother.

"Of course I took a lot of flak in the Governor's Council and in
the papers but finally the appointment went through. The day after it was announced in the *Globe*, Bulger came in again. 'I'm really
sorry about this, Governor, but I don't think I want that judgeship
after all.' He explained that Kevin Harrington was going to resign
from being President of the Senate and Bulger thought he would
have a shot at it. 'I'm really embarrassed about this, Governor. I
know the *Globe* has given you quite a beating.' I assured him it was
okay. We'd take a little more flak but it would pass.

"Well, the day after he was elected President of the Senate one
of my bills came up for a vote and Bulger squashed it. One of his
friends came up to Billy and said, 'Fer Christ sake, Bulger, how
could you possibly do that after all that Sargent has done for you in
the past few weeks? Bulger winked and said, 'Yeah, well, what has
he done for me today?' I thought I'd bust a gut I laughed so hard.
Of course Bulger wanted me to hear it."

But most of all I remember my father's stories about when he
and my mother moved to Cape Cod. "Jess and I moved to the
Cape right after the war," he would recall. "I had decided I didn't
want to go back to the city to be an architect so we built a little
camp across the river on Pleasant Bay. I was doing some carpentry
then and running a charter boat out of Rock Harbor. One day
Billy Gould and I were up on the roof shingling the camp when his
father came by. Willis Gould had been an old market hunter.
Those were the guys who used to shoot ducks and geese commer-
cially for the markets in Boston and New York.

"'Now goddammit,' Willis started in, 'you boys know the
gunnin' season's ovah and you shouldn't have that shotgun up
theah.' Just then we heard, *aahonk, aahonk*. It was a whole skein of
geese flyin' low over the bay. Willis stopped in mid-sentence.
'Gimme the gun, gimme the gun!' he said, and knocked down two
stragglers flyin' overhead.

"Every Sunday Jess and I used to walk along the shore to a beautiful little house set on a bluff above Pleasant Bay. We always talked about how much we'd like to live there one day. Well, this Sunday the owners happened to invite us in. It seemed there had been a death in the family and they were about ready to put the house up for sale and move to Texas. To make a long story short, we agreed to buy the house that very afternoon. I think we paid twelve thousand dollars for the house and five acres, plus another two hundred bucks for the boats and furniture."

Well, that really gets us to the beginning of this book. In my earlier books I've written mostly about animals, nature and science. Occasionally I'd throw in a story or two about people. But I tended to avoid such stories because I knew how difficult they were to write. When I did write about people I relied on my father's ability to recognize a good story. No writer had a better source, no anthropologist a better informant. He'd pick up the stories out on the bay, in the boat, or by swapping tales at the Goose Hummock Shop.

In this book I've also told a few stories about people. I have tried to tell them as accurately and truthfully as I was able to. I was not always on the beach, or in the bar, and certainly not in the beds where some of these conversations took place. But good stories by nature are more truthful than factual. My father always knew that. I hope I have done a good enough job so that if he were to read them he would say, "You know, that was a damn good story."

W. S.

Introduction

Reflections Over New Jersey

Shortly after finishing the manuscript for this book, I flew up the East Coast from Florida to Boston. It was on one of those clear cobalt blue days when the shore stands out starkly against the deep green of the ocean. Below me were the streets of Atlantic City made famous by the game of Monopoly. I could almost pick out Boardwalk and Park Place, Ventnor and St. Charles. I could see the utilities that supported the once-thriving resort and the railroads that had put it on the map.

I could also see seawalls that outlined the shore and a lattice-work of groins and jetties that poked into the Atlantic like tooth-picks skewering a plateful of canapes. The seawalls were the real-life property owners' favorite method of preventing their expensive hotels from toppling into the Atlantic. The groins were their favorite method of trapping the sand washed away because of the seawalls.

These skeletal artifacts were the remains of an era when Atlantic City reigned as the queen of American resorts, when incoming trains disgorged presidents and dignitaries, socialites and starlets. It was the place where John Philip Sousa sent chills down the spines of summer audiences and Jack Dempsey jogged along the beach to prepare for his bout with Gene Tunney.

But today the beach is gone, the hotels are boarded up and the city is used by coastal geologists as a textbook example of the futility of attempting to battle the effects of sea level rise.

From an altitude of 3,000 feet it is easy to see the folly of building on this fragile shore. It is easy to condemn the greed and short-sightedness of rapacious investors and the mendacity of government officials. But, from 3,000 feet, it is also easy to miss the human story; to disregard our deep-seated instinct to defend our homes and to forget the overwhelming sense of loss we feel, when we lose a battle to keep a home. Each house, each dock, each deck and boat below me had witnessed human stories. Some people had won brief battles with the ocean; others had to endure tragedies that would reverberate through their families for generations.

It was to investigate this kind of story that I spent seven years walking the beaches, studying the marshes and talking to the inhabitants of Chatham, Massachusetts. I spoke with fishermen, lawyers, homeowners and scientists as they struggled to come to grips with the effects of sea level rise and global warming—considered to be the two raciest horses of the coming apocalypse.

The story has protagonists, antagonists, drama and action. I've talked to hundreds of people facing the prospects of losing their homes, their jobs, their livelihoods. I've seen them make good choices and bad, wise decisions and dumb ones.

During all these encounters, I have never met a person whom I would describe as thoroughly bad, or one who acted even faintly like the self-centered, fat little capitalist who dashes around the Monopoly board. What I did meet were people who were doing their best to figure out how to live in a changing world. It is a story singularly devoid of heroes in pure white hats and villains in pure black ones.

I have tried to tell the story through the eyes of the people involved. I have also relied on newspaper articles, scientific papers, and court transcripts to supplement their memories of past events.

I apologize to some of my informants who had to put up with my innumerable follow-up calls. I am sure they often questioned my abilities if not my sanity when, after finishing an hour-long conversation about a fine point of geology or a past congressional hearing, I would call back to check on the color of their telephone or the depth of their keel. During the course of making these calls I made a discovery that probably should have been obvious, but was entirely new to me. Often when I called a fisherman he would be at sea and I would end up having long talks with his wife. It soon dawned on me that women remember the personal context of past events far better than men.

Gradually, I made a point of calling the wives of scientists, lawyers and homeowners to ask about their memories of particular incidents. I am sure many of my informants were curious about why I wanted to speak to their wives, but I had discovered that when men discuss their role in resolving a particularly thorny issue, they often have the unfortunate habit of forgetting the whereabouts of their wives or the names of their children.

Of the hundreds of stories I could have used to tell the story of Chatham I have selected only a few, including my own, that I hope will be representative of the many separate episodes. Another writer might have selected other stories and written an entirely different book. But I hope the way I have told the story captures the essence of both the human drama and the scientific controversies surrounding the incidents in Chatham, and the world.

I was fortunate that the scientific story of Chatham happened to span one of the most dynamic eras the East Coast has ever faced. The conclusion of the 1980s saw the hottest years on record and a peak of concern about global warming and sea level rise.

The 1990s were ushered in with dramatic cooling and a succession of severe storms that caused unprecedented erosion. As the story of Chatham unfolded, scientists were trying to unravel the potential causes and implications of these new patterns of weather and climate.

I have tried to recreate the mood and language of the scientists' conversations as they tried to deal with the effects of global warming and sea level rise. In my attempt to capture the thrust of some of these conversations, I have used literary license. The two chapters that describe what happened in a Chatham bar are examples. The conversations I relate never took place in one room at one time but are instead a distillation of similar conversations I have had with many different scientists.

As a science writer, I have discovered that often the most effective way to find out what a scientist really thinks is to get him out of the lab for an "over a beer" off-the-record conversation. It is, of course, a strategy that political writers have used for years. If anyone takes issue with the veracity of my conversations in the Squire, my only defense is that I am sure they are remembered at least as accurately as most other conversations that have taken place in this fine establishment.

Finally, I must add a personal note that undoubtedly has some bearing on this book. I grew up in the house that my parents bought that overlooks Cape Cod's Pleasant Bay. Today I live in Woods Hole, in a house that is half a mile inland. Many people might consider that I still live on the water, but I know I do not. Most days I never even see the ocean. Anyone who has ever had the privilege of living on the water knows how different it is from being an inlander—one is like visiting Paris, the other is like being a Parisian.

When you live on the water the moods of the ocean shape your life. You feel the power of nature when an Atlantic storm pounds the shore. You marvel at the ocean's beauty when her waters reflect a shimmering trail of moonlight on a quiet evening in June. You alter your day's activities to take a walk after spotting seals on a rock or to go fishing after seeing terns working over bait at dusk.

There is a world of difference between living on the water and going to the beach. The first gives you an abiding sense of place; the other, a pleasant afternoon.

I would not be doing what I am doing today or writing the books I write if I had not grown up on the water. Many other authors have also been profoundly affected by their surroundings. Could Thoreau have written *Walden* from downtown Concord or

Henry Beston have written *The Outermost House* from the center of Eastham? Living on the water launched my career and stamped its imprint on my life forever. If that experience has shaped this book, I acknowledge it, without hesitation, as a stated bias.

William Sargent
Woods Hole, Massachusetts

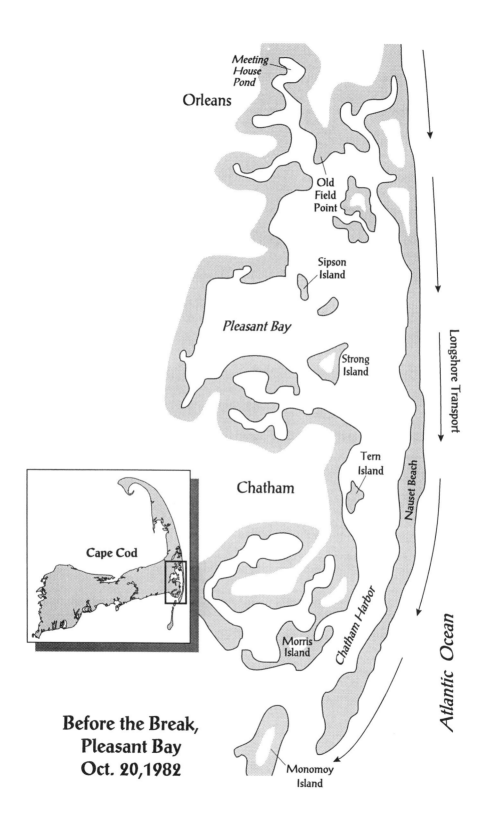

Meeting
House
Pond

Orleans

Old
Field
Point

Sipson
Island

Pleasant Bay

Strong
Island

Tern
Island

Chatham

Longshore Transport

Nauset Beach

Chatham Harbor

Atlantic Ocean

Cape Cod

Morris
Island

**Before the Break,
Pleasant Bay
Oct. 20,1982**

Monomoy
Island

PART I

The First Year
1987

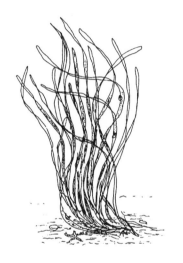

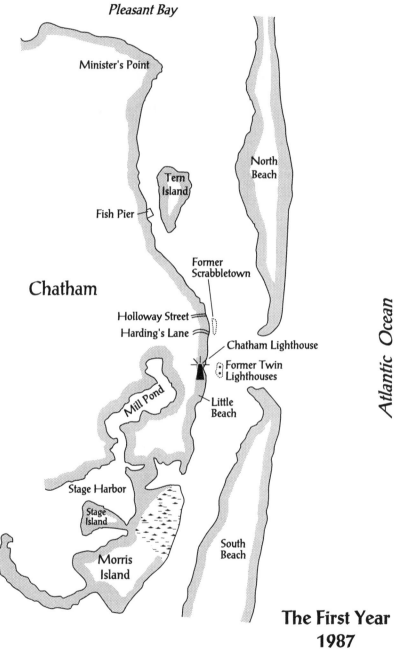

Pleasant Bay

Minister's Point

North
Beach

Tern
Island

Fish Pier

Former
Scrabbletown

Chatham

Holloway Street

Harding's Lane

Chatham Lighthouse

Former Twin
Lighthouses

Mill Pond

Little
Beach

Atlantic Ocean

Stage Harbor

Stage
Island

Morris
Island

South
Beach

**The First Year
1987**

CHAPTER 1

The Break

January 2, 1987

Everything's gonna be different in the mornin'.

—*A Fisherman, January 2, 1987*

I hate northeasters more than any other kind of storm. Hurricanes bore in from the tropics then dissipate quickly over land. A northeaster just sits off the coast like a deep, gray, winter depression that won't go away. A hurricane you want to name and remember, a northeaster you want to forget.

The year's first northeaster hit on January 2nd. Most of the fishermen were in port for the holidays but were eager to get back to sea to make their next boat payment. But the sun had disappeared behind rolling gray rain clouds; the wind freshened from the northeast and large, fat raindrops started to splatter on the sandy soil of Cape Cod. By noon the storm was lashing the low, gray houses that huddle around the center of Cape Cod's towns. It pelted the empty, white, summer houses that stood so proudly on her high cliffs.

Soon the deep groans of foghorns started to sound from one end of Cape Cod to the other. I listened to the Martha's Vineyard ferry as it paused to catch the rapid red and white flashes of Nobska Light, before slowly retracing her wake back to Woods Hole.

In Chatham, pickup trucks converged on the Fish Pier as captains made one last check on their boats before heading home, or to the nearest bar. Northeasters do that, too. They lock some men at home, hopeless, helpless and in despair—they drive others to the nearest bar. By evening the storm was raging. The twin beacons of the Chatham Lighthouse cut through heavy squalls that were gusting over the whitecapped harbor.

The storm had started when a front of arctic air had collided with the moist tropical air that rises off the Gulf Stream east of Cape Hatteras. Winds had spilled out of the eastern high pressure cell and swirled into the westerly low. Soon the cyclone of winds had broken off and gained its own identity.

The new northeaster had picked up momentum and moved slowly up the coast. Off Cape Cod, it had stalled and its winds started to lash the shore for several days. Rain turned to sleet, then hail, then snow. Winds piled up waves 15 feet high and the storm's low pressure had raised the sea level a foot above normal. Waves undercut the foundation of Nantucket's Great Point Lighthouse.

Along the outer arm of Cape Cod, the storm piled up water at the base of the high cliffs of Provincetown, Truro, Wellfleet and Eastham. Waves pounded the cliffs and dragged heavy boulders in their backwash. One wave would tear a two-ton boulder from the cliff, the next would catapult it back again. Wave after wave tore up the fragile striations of sand, gravel, clay and boulders laid down by the Wisconsin glacier thousands of years before.

Further south, breakers flattened the low sand dunes and breached North Beach, the fragile barrier that protects Chatham Harbor. Through the break, water rushed into the harbor and Pleasant Bay. The bay was already several feet higher than usual. High tide came on top of a foot-high storm surge that had been lifted by the storm's low pressure core. Tons of sand and water were flowing freely into the bay, but the real damage would come after the storm had passed.

As the winds died and the tide turned, the system gradually relaxed. The water that had been forced into Pleasant Bay had to go somewhere. The nearest outlet was the overwash area between the dunes. The water rushed back through the new breach toward the Atlantic. It scoured a foot-deep channel down the center of the breach and started to deposit a spreading tongue of sand onto the floor of the Atlantic Ocean.

With each successive tide the breach was scoured wider. Soon an eighteen-foot stream meandered back and forth across the overwash area that had grown as broad as a football field. It's waters, at only a foot deep, were still wadeable, but they flowed unimpeded from the Atlantic Ocean to Chatham Harbor—a new inlet had been born . . .

In the middle of the night a fisherman belched, broke wind and rolled on his back. His wife's sharp elbow brought him to full consciousness and he listened intently to the night. There was a new and urgent sound on the distant beach.

"Sompin's happen'd out there. I can hear it in the surf. Don't know what it is. But everything's gonna be different in the mornin'."

His wife snored peacefully beside him.

CHAPTER 2

Chatham, Massachusetts

February 1987

> A veritable town at sea, lying farther oceanward
> than any other town in the United States.
>
> —*George "Chart" Eldridge, 19th-century,*
> *Chatham native and founder*
> *of* Eldridge's Tide Charts

It was the end of February before I was able to drive from Woods Hole to Chatham to see the changes caused by the new inlet. Another northeaster, the third of the season, was splattering Cape Cod with more snow and freezing rain.

I drove my car slowly down the mid-Cape highway, devoid of oncoming vehicles. My headlights could barely pierce through the heavy, wet snowflakes that swirled dizzily out of the darkness that surrounded my lonely automobile. All that could be heard was the quiet swishing of my wipers as they attempted to alter the course of slushy snow that slithered back and forth across the windshield. The defroster could barely maintain a small opening in the rapidly accumulating snow and ice. I crouched low in my seat in order to peer through the tiny hole, as a cold, dull ache grew steadily in the base of my spine.

I had spent the last hour searching in vain for exit signs among the scraggly branches of the low pines that huddle atop the banks

9

of the barren highway. Finally, a small sign loomed out of the swirling snow. I eased off the highway and drove down snowbound
streets to the deserted center of Orleans. I had driven several miles
out of my way in order to enter Chatham from the north on Route
28. I did it partly out of habit, because I had grown up in Orleans,
and partly by design. My detour would allow me to see the effects
of the new breach on Pleasant Bay, which runs parallel to Route 28
from Orleans to Chatham.

Fishermen say that you can tell the difference between Orleans
and Chatham by color. From offshore, Orleans is green with trees
and small cedar-shingled homes. Chatham is white from the summer mansions that dominate her shores. Even in the snow I could
see the difference as I crossed the town line.

The white homes are owned mostly by wealthy summer resi-
dents. However, even the least perceptive visitor soon realizes that
their owners are not the true aristocracy of Cape Cod. That distinction belongs to the Nickersons, Eldridges, Snows and Doanes—the
first families that settled here nine generations ago. Perhaps more
than any other Cape Cod town, Chatham retains it's historic roots.
Eldridges and Nickersons still own the businesses, run for the office
of selectman and serve on local boards.

In fact, you can navigate through much of this part of the
world using only old family establishments as landmarks. Old timers do it unintentionally when giving directions. "Start out at
Nickerson's lumber, come down Eldredge Parkway, then keep
right past Nickerson funeral home. When ya come ta Josh
Nickerson's place you'll be in Chatham."

Come to think of it, you can pretty much navigate through life
on Cape Cod using only Nickerson establishments. "Start out at
the Nickerson room of the hospital, buy a house at Nickerson's
Real Estate, add an addition with Nickerson lumber, and keep going as long as ya can 'til ya end up in the Nickerson funeral home."

My musings had carried me through the cold night to the
Chatham Fish Pier, the spiritual heart of the town. The Fish Pier is
where Chatham officially greets the Atlantic Ocean. The harbor is

home to one of the East Coast's most lucrative fishing fleets. With only 70 boats, it ranks lower than New Bedford and Gloucester in landings but often scores ahead of Boston in the value of it's catch.

In all, Chatham's small fleet supports over 500 people, including fishermen, shuckers, net hangers and wholesalers. They bring in $25 million dollars, not small change for a town with only 6,900 year-round residents. Needless to say, it is only fitting that Nickerson's Fish & Lobster dominates the pier.

The reason that Chatham's fisheries are so lucrative is that the town juts into the Atlantic farther than any other major fishing port. It sticks out 40 miles east of the mainland and is surrounded by the Atlantic, Nantucket Sound, and Pleasant Bay. With 66 miles of shoreline and only six-and-a-half square miles of upland, Chatham's very existence is inextricably bound to the Atlantic Ocean.

Beyond the fish pier the road rises to the Chatham Lighthouse. If the fish pier is the heart of Chatham, the lighthouse is her eyes. It is to the Chatham Lighthouse that people go to fully grasp the tenuous relationship of Chatham and the sea. It was to the lighthouse that I went so that I might see the effects of the new inlet.

The double beacon of the Chatham Light arced overhead. The driving snow almost obscured the few cars huddled at the end of the parking lot. Their occupants peered through steamed-up windows but saw little. The harbor, the breach and the Atlantic Ocean were all lost in the whirling snow. The only sounds were the groan of the foghorn and the ominous roar of the ocean that lay somewhere offshore, unseen and threatening.

By daylight, people go to the Chatham Light in all seasons to gaze at North Beach, the narrow barrier of sand that used to protect the mainland. Since January 2nd, however, that barrier had been broken. The full force of the Atlantic Ocean could now surge through the ever-widening inlet to attack the bank below the lighthouse and the homes on Watch Hill, Holway Street, Andrew Harding's Lane and Little Beach. Like a dike in the Netherlands, North Beach had become the focal point of people's concern about the future.

Below the lighthouse a lone white obelisk reminds viewers of the dangers of these waters. On March 11, 1902, two schooner barges ran aground on Shovelful Shoal a mile off of Monomoy Island. Surfmen from the Monomoy Life Saving Station rowed the crew members to safety but their owners ordered them back aboard to jettison the cargo, and save the ships.

A few days later the winds picked up and the bargers flew a distress signal. Seven members of the Monomoy Station rowed out to investigate. The bargers demanded to be taken off, even though they were in little danger. The surfmen took aboard the bargers and Captain Eldredge ordered them to lie down in the bottom of the boat. When the first waves hit, however, the bargers panicked, and grabbed the oarsmen around the neck. Unable to row, the surfmen lost control and the small boat capsized in mountainous swells.

Exhausted from rowing, the surfmen could do little except cling to the wreckage of the overturned boat. One by one the men lost strength and dropped off. Surfman Chase. Surfman Nickerson. Surfman Small. Surfman Kendrick. Surfman Foye and Captain Eldredge. Surfman Rogers was the last, moaning that he could not go on, as he slipped below the waves. Only Surfman Ellis managed to hang on long enough to be saved by the equally heroic efforts of Elmer Mayo, captain of the schooner *Fitzpatrick*.

The barges stayed safely on the shoals for several more days. The official report of the incident blamed "the needless panic of the barge crew and the high sense of duty of the surfmen that would not permit them to turn their backs upon a signal of distress." It became known as the Monomoy Disaster, the worst tragedy in the long history of the United States Life Saving Service.

At the base of the obelisk, an excerpt from a poem by Tennyson commemorates the local heroes and the ever-present danger of navigating the Chatham inlet:

> *Twilight and evening bell,*
> *And after that the dark!*

And may there by no sadness of farewell,
When I embark;

For tho' from out our bourne of Time and Place
The Flood may bear me far,
I hope to see my Pilot face to face
When I have crost the bar.

But it was time to leave the lonely overlook to seek out some warmth and human companionship. About the only place to find such things was at the Squire, a bar and restaurant, the social center of late-night Chatham.

By day, the Squire is jam packed with the shopkeepers, salesmen, firemen, cops and local officials who keep the town running. By night, the Squire is the domain of lobstermen, cod fishermen, gill netters, shuckers and shellfishermen. Once the dominant economic force on Cape Cod, fishermen have lost their economic power to real estate and tourism, but they still maintain their mythic hold on the psyche of Cape Cod.

As I swung open the double doors of the Squire, I was assailed by the warm, rich odor of beer, quohog chowder and hot pastrami sandwiches. The dark walls of the bar were lined with old Cape Cod names on carved wooden signs. There were the Nickersons who bought (some say stole) most of Chatham from the Indians; the Norgeots, who came from St. Pierre and Miquelan to build the French cable station that once was the only direct connection between the United States and Europe; the Eldredges with an "e," who are remembered because they stole horses, and the Eldridges with an "i," who are remembered because they did not. Such things are not soon forgotten on Cape Cod.

Along the walls were paintings and photographs of Stage Harbor and the fish pier. A large sepia mural depicted one of Chatham's old life saving stations, complete with a horsedrawn surf boat at the ready to rescue one of the hundreds of vessels that have foundered off this treacherous coast. Many of the people seated in the restaurant could still remember the sudden absence of a high school classmate when her father was lost as his boat foundered, "tryin' to crost the outer bar."

That night, all conversation revolved around the breach. Several fishermen who were huddled below the incongruous totem of a stuffed sailfish, answered my questions. "Yup, most of the guys are going through 'er now. I hear Mark Farnham made it through yesterday with his 40 footah. Dragged his ass, but made it through, slicker'n shit."

"It's still pretty damn dangerous. A ten-foot wave caught Chris Armstrong's boat a few days ago. Lifted 'er into the air and dumped 'er right on a sandbar, cracked her god damn keel. Hell, he'll be out most of the season and he just finished rebuild'n his engine."

"Will it help any?"

"If she don't close up it'll help a lot. Maybe knock an hour and a half off the trip to the banks. That'll help with fuel bills some, too. Nick Brown figures he'll save couple thousand a year just on fuel."

Two other fishermen were sitting at a table beneath Eldridge's 1890 chart of the old breach. One joined the conversation.

"She'll close up sur'n hell. Remember '78, Nauset broke through in 13 places and they all fill'd in in less than a week."

"Yeah, but Monomoy stayed open. I went through there all last summer when I was get'n horseshoe crabs."

"You never did tell anyone how much you got paid for those crabs. What was it they were doin with 'em?"

"Bleed'n 'em for drugs, and I'm still not going to tell you."

"Did I hear the tides in Meeting House Pond are a foot higher?"

"Yup. Will Case up ta Nauset Marine is gonna put a tide gauge in for that guy Giese from Woods Hole."

"Jesus Christ, a foot higher. That's goin' to raise some eyebrows. I hear some of the summer people are already hiring lawyers to build seawalls."

The last revelation gave me some pause. My parents' house was on a bluff overlooking Pleasant Bay. If the tides were a foot higher in Meeting House Pond, what were they doing to the bank below our house?

Outside, the snow had turned to a driving rain. I left the bar and drove back to East Orleans and the long driveway that led to our darkened home.

CHAPTER 3

Beginnings

March 21, 1987

Glacial epochs are great things, but they are
vague, vague.

—*Mark Twain*

I timed my next visit to Chatham to coincide with the first day
of spring. I did it for several reasons. The opening of the new
inlet had ushered in a new era. It was the beginning of an elegant
experiment. Pleasant Bay had become a new ecological laboratory
that could be used to investigate the effects of sea level rise. The
inlet had increased the tidal range of Pleasant Bay by a foot. The
bay was now six inches higher at high tide and six inches lower at
low tide. I was curious to see how these new conditions would
affect the plants, animals and geology of Pleasant Bay.

As with any experiment, the best way to begin is to record
what the area is like before the experiment starts. Besides, the first
day of spring just seemed like the most poetic time to start such a
project. The day lived up to expectations. The new spring sun cast
it's first warm glow on Chatham. It sparkled on the cold, green
waters of the Atlantic that were streaked white with the foaming
wakes of incoming breakers—breakers that now passed unhin-
dered through the mile-wide inlet, which had grown much faster
than anyone had anticipated.

Nauset Beach reflected the sun. It stretched eight miles north to Orleans, while Monomoy Island reached six miles south toward Nantucket. Behind the broken barrier of Nauset Beach, the shallow waters of Pleasant Bay rippled in the new spring sun. They lapped the sculpted bluffs of her western shore and nourished the hidden roots of eelgrass that would soon thrive in the deeper holes and channels of the bay. Around the perimeter of Pleasant Bay, tiny shoots of marsh grass were poised beneath the stubble of last year's growth. Soon they would emerge to paint the marsh green with their abundance.

Pleasant Bay did not always look this way. It was born in an icy crucible between two lobes of the glacier that made Cape Cod. Twenty-one thousand years ago the world was locked in the Wisconsin stage of the last ice age. The sea level was 300 feet lower and the coast of the Atlantic Ocean was 180 miles to the east, at the edge of a frozen continental plain. The Laurentide ice sheet had pushed as far south as Martha's Vineyard and Nantucket. Today the ridge of rocks, sand and gravel that make up the islands' terminal moraines mark the furthest incursion of the ice sheet.

Eighteen thousand years ago the world was rapidly emerging from the ice age. The sea level was rising almost 3 feet every 50 years, and all the glacial activities that made Cape Cod took less than a thousand years. The glacier that had scraped across Canada and northern New England in it's earlier advance had stalled over Cape Cod. It was in an erratic phased retreat. It was still expanding from it's mile-thick center, incorporating boulders, rocks and gravel within it's icy grasp. However, the lobes of ice that made up the glacier's leading edge were simultaneously retreating under the warming influence of the Gulf Stream. In the wake of their retreat, the forward lobes left huge blocks of ice that formed the kettle ponds and deep holes of Pleasant Bay.

The glaciers also left behind end moraines, long ridges of rock, sand and gravel that were pushed up by minor advances of the stalled glacier. The most familiar of these moraines is the one created by the Cape Cod Bay Lobe. This moraine forms the gently curving ridge of glacial material that forms the distinctive "bare and bended arm of Cape Cod" that Thoreau wrote about in 1855.

Today, motorists are most likely to experience the Cape Cod Bay end moraine as they sit stalled in traffic on Route 6. The mid-Cape highway was built intentionally on the natural high berm that was pushed up by the icy lobe. Five thousand years from now this high spine of Cape Cod might be all that remains of today's sandy peninsula.

The South Channel Lobe of the Wisconsin glacier is less well known because most of it's end moraine lies underwater. The South Channel Lobe met the Cape Cod Bay Lobe at the head of Pleasant Bay and bulged southeast toward modern-day Georges Bank.

The intersection of these two walls of dirty, rock-strewn ice was the crucible that formed Pleasant Bay, Orleans and Chatham. Meeting House Pond was a drowned kettle pond at the head of the bay and Meeting House River was a torrent of white meltwater that rushed from beneath the two lobes of the icy blue glacier.

The meltwater river ran into a depression left by the weight of a former fragment of the South Channel Lobe; there it formed a glacial lake. Perhaps another river flowed out of the lake toward the Atlantic Ocean off Nantucket or meandered eastward across the frozen continental plain to the distant ocean shore.

The only evidence of the South Channel Lobe that remains above water is the northwest shore of Pleasant Bay, Sampson's Island, Hog Island and a string of erratic boulders that stretch toward Nauset Beach. The outer edge of the moraine lies beneath the Atlantic Ocean and makes up Georges Bank, one of the modern world's richest fishing grounds. However, ten thousand years ago Georges Bank was one of the world's richest hunting grounds. It had become a fifty-mile-long island with remnant herds of mastodon and woolly mammoth. Paleo-Indians may have hidden in it's groves of spruce and fir trees and may have contributed toward hunting the magnificent animals until they reached the brink of extinction.

Nine thousand years ago the ocean continued to rise, but at a slower pace. It finally washed over Georges Bank, perhaps drowning the early Indians and the mastodon. Certainly the Indians of this region saw dramatic incursions of the ocean during their lifetimes.

The drowning of Georges Bank had severe consequences for Cape Cod. Now the irregular pile of rubble left by the receding glacier had to contend with the prevailing easterly winds that raged against the outer arm of the Cape. Without the interference of the old island, waves from the southeast pounded directly against the rugged headlands of the outer Cape, initiating a period of intense erosion and beach building that formed proto-Nauset Beach and continues today.

The waves that crashed against the headlands also piled up water against the shore. The water had to go somewhere, so it formed longshore currents that flowed in the rips and runnels that parallel the shore. These longshore currents are the same ones that carry swimmers lazily along the beach. However, when the currents encounter a break in the sandbars, they turn directly out to sea, becoming the ill-named but treacherous rip tides that have swept countless swimmers to their death.

The longshore currents also played a more prosaic role. They worked in collusion with the waves to move sand along the shore. Each wave stirred up sand from the beach, suspended it in water, and dragged it off the beach. Then the sand was carried parallel to the shore by the longshore current. The next wave redeposited the sediment back on the beach as a delicate little filigree of sand grains. However, each grain had made a long elegant loop that moved it downstream from where it had been before. Even on the calmest day in summer, the beach moves right past swimmers who are unaware of the subtle migration occurring before their naked feet.

The process makes Nauset Beach one of the most powerful sand transport systems on the planet. The south end of Nauset Beach grows 200 to 300 feet a year, or about a mile every ten years. We can see the process from space. If we discovered something like this happening on another planet, it would receive worldwide attention.

Five thousand years ago sea level was still 16 to 18 feet lower than today, but it was rising about six inches every fifty years, about the same rate as occurs today on Cape Cod. Pleasant Bay looked much like its modern counterpart, but it was three times larger and almost six miles wider.

Nauset Beach was about 14 miles long, extending most of the way to Nantucket. As the sea level continued to rise, however, sand was eroded off the ocean face of the beach. During storms, sand would also wash over the beach and be deposited on the Pleasant Bay side of Nauset. In effect, the beach would roll over like the track on a military tank. Sometimes Nauset Beach would migrate landward as much as 20 feet in a single storm, but on average it migrated westward 5 to 10 feet annually. Thus Pleasant Bay grew narrower every year. Today it is only 3 miles wide.

Eventually, the roll over of the southern portion of Nauset Beach outpaced the roll over of the northern portion. The beach broke in two, creating Monomoy Island. Today Monomoy continues to roll over more rapidly than Nauset, leaving the two beaches curiously offset. In the roll over sweepstakes, Monomoy Island leads Nauset Beach by a half mile at the turn.

Behind Nauset Beach the waters of Pleasant Bay were calm and protected, ideal for shorebirds and waterfowl. At about the time that Hammurabi was codifying the laws in Babylon and the Pharaohs were building their pyramids, a duck flew up from the south and alighted in Pleasant Bay. A few marsh grass seeds dropped off his feet or were deposited in his feces. They sprouted and the marsh started growing, keeping pace with the rising sea level. If you were to drill down through the marsh today you would recover a core of peat 18 feet thick, an indication of how much the sea level has risen during the last 5,000 years.

Two thousand years ago Pleasant Bay looked very much like it does today. The sea was still 6 feet lower and Pleasant Bay was

probably two miles wider. Nauset Beach was continuing to migrate west, rolling over the new marsh. Today swimmers can see
the results of this roll over after major storms. Waves often uncover
a narrow ridge of peat in the surfline at the edge of the outer
beach. The peat is what remains of the marsh that once grew on
the Pleasant Bay side of Nauset Beach but was rolled over by the
migrating sand.

Perhaps the most dramatic example of this continual process of
roll over occurred when the *Sparrowhawk* ran aground inside
Pleasant Bay in 1626. Governor Bradford, the first governor of the
Massachusetts Bay Colony, duly recorded the disaster and it was
gradually forgotten. However, in 1863 a winter storm revealed the
bleached ribs of the *Sparrowhawk* on the ocean side of Nauset
Beach. During the intervening 237 years, Nauset Beach had migrated west 1,000 feet, rolling over the sepulchral remains of the
forgotten wreck.

Today, the process continues; the sea level rises, overwashes
occur and Nauset Beach retreats into Pleasant Bay. It may appear
like a calamity when overwashes and breaches occur, but in the
long run the westward retreat of Nauset Beach maintains the barrier between the rising ocean, the bay and the fragile uplands.
Without this landward migration, the rising sea would quickly
drown Nauset Beach, exposing the fragile glacial remains of Cape
Cod to the full fury of the Atlantic Ocean.

In ancient China, noblemen were taught the lesson of the
oak and the bamboo. The mighty oak tree tried to fight the wind
but it was snapped in two by a storm, while the graceful bamboo
bent before the wind and survived. Like the bamboo, Nauset
Beach bends to the rising sea and survives by retreating landward, while the rock-bound coast of Maine fights the sea and is
toppled.

Four thousand years from now, much of Cape Cod may still be
present as a long, thin barrier beach that connects Cape Cod to
Nantucket, and Nantucket to Martha's Vineyard and the mainland.
It will be a mature barrier beach system protecting thousands of

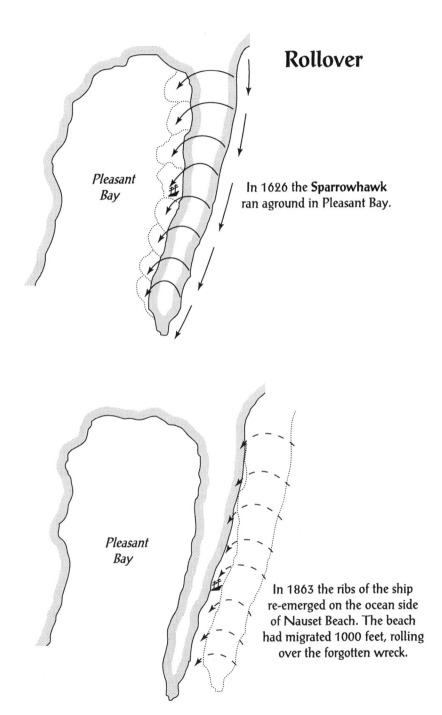

Rollover

In 1626 the **Sparrowhawk** ran aground in Pleasant Bay.

Pleasant Bay

Pleasant Bay

In 1863 the ribs of the ship re-emerged on the ocean side of Nauset Beach. The beach had migrated 1000 feet, rolling over the forgotten wreck.

acres of marsh grass. In that respect, it will be similar to the barrier beaches that protect the present-day shores of the Carolinas.

When it happens, future Cape Codders may feel some comfort that their barrier beaches have saved them from the fate of the cities and rock-bound coasts of New England that will have long since disappeared beneath the sea.

CHAPTER 4

The Nauset Cycle

April 1987

Ye Scrabbletowners, ye Chatham wreckers, Git
oout with your brick in your stocking.

—*Rudyard Kipling,* Captains Courageous

Winter does not release its grasp easily on Cape Cod. It still lingers in the cold gray waters of the Atlantic, emerging at will to cool the fragile peninsula. Easterly winds carry the chill across the landscape in gossamer shrouds of fog, rain and drizzle.

Offshore, ten-foot waves pound the sandy flanks of Nauset Beach. Raging winds tear the foaming tops off incoming breakers and blow them sideways toward the shore. On such a day, the towering surf dwarfed three small figures who were squatting in the sand at the edge of the roiling sea.

Graham Giese huddled against the wind, trying in vain to write figures in the rain-soaked notebook balanced on his left knee. Freezing rain splattered his unprotected face and salt spray pelted his sopping parka. The Woods Hole geologist had to yell to be heard over the steady roar of the crashing surf. The faint remnant of his Southern drawl melded incongruously with the clipped New England cadence of Richard Hiscock, Chatham's assistant harbormaster. Despite the weather, there was no place the two men would rather be and no place else they should be.

Dr. Giese peered up at Hiscock. "Okay, Richard. Let's use the wrackline from last night's tide to establish mean high water. I'll hold this end and you run the tape out to the next stake."

Hiscock crouched low and strung out the tape from the wrackline of flotsam and jetsam that delineates the furthest incursion of the tide toward a tall stake, stuck in the sand at the edge of the duneline. He shouted back through the driving rain, "209 feet."

"Holy cow, that's over 200 feet we've lost since last month and it looks like at least nine more stakes have washed out down the beach. Come on back."

Hiscock paused to catch his breath. "One of them is out there now, drifting just behind the first breakers. Could become a risk to navigation."

"Yeah, the longshore current's got it. Maybe we should switch to aluminum stakes. They'd sink as soon as they got swept off the beach. You know, we've lost close to a football field worth of sand off South Beach since we were out here last month."

"Incredible."

"Yup, pretty impressive, one of the most powerful sand transport systems in the world."

Dick Miller, a volunteer with the Chatham Waterways Committee, drew in the strings on his parka hood. "How does the sand get onto the beach in the first place?"

"See those little ridges in the wrackline? Well, each wave leaves a tiny ridge of sand like that. In the aggregate, all those ridges build up the beach. They make Nauset Beach grow about a mile every ten years. That's about a dozen football fields every year."

"And now since the inlet we're losing a football field worth of sand to the inlet every month. Where's it goin'?"

"It's either getting swept into the harbor to form shoals or it's getting deposited at sea on the ebb tide delta."

"So the beach south of the inlet isn't getting sand any more?"

"That's right, South Beach is getting starved. In twenty or thirty years it will have mostly washed away and migrated over to the mainland. When that happens, North Beach will start growing rapidly again. In about 140 years North Beach will have grown to where it was before, and Pleasant Bay will have to find a new inlet again."

"So the cycle will start all over. Amazing, but didja ever expect it to go this fast?"

"I expected it, but I still can't believe it."

"Boy, it's gett'n cold out here. At least you didn't jump into the water before we hit the beach like last time."

Graham's absent-minded enthusiasm was a standing joke among the three men. They liked working together. Dr. Giese's enthusiasm and easy-going style were infectious. Each trip became another animated lesson in coastal geology.

Hiscock already knew a lot about Dr. Giese's work. In 1976 his mother had asked Graham to prepare a history of Nauset Beach for the Chatham Conservation Commission. Since Giese had gone to surveying school while serving in the army, it only made sense that he would turn to old charts to re-create the history of the beach.

Consulting the charts was the right approach. Giese found accurate charts that spanned the last two hundred years of changes in the system. They revealed the regular patterns of the Nauset Beach Cycle.

Giese noticed that it took 140 years for Nauset Beach to grow long enough that it constricted the flow of water in and out of Pleasant Bay. The tides were then delayed and the upper bay became stagnant. The constriction of the throat of the bay built up hydraulic pressure behind the narrow beach. Water in the bay then wanted to get out and, when a washover provided a detour, the pent-up water quickly gouged a new exit and a new inlet was born.

By nature and training Giese was a conservative man, so he had rounded off his estimate of the Nauset Beach Cycle to 150 years. In fact, the new inlet broke through 140 years and 2 days after the old inlet had formed in 1846. If Giese had stuck to his guns, he would have been only two days off, not a bad margin of error for a

140-year cycle. If only earthquake forecasters and weathermen could do as well.

Hiscock paused to stretch and pointed to the mainland. "You know, the last time the inlet opened, Chatham lost the old twin lighthouses."

"I read about that. Erosion undercut one tower and it pitchpoled down the bank. Smashed into a thousand pieces on the beach 80 feet below. Second one went two years later."

"Yup, lost a whole village too."

"Really? Where was that?" asked Miller.

"Old timers called it Scrabbletown. It was nestled into the bluff over there just to the right of today's lighthouse."

"What was there?"

"Oh, not too much really. I guess mooncussers and wreckers in the old days."

"Those were the guys who used to lure ships to their doom with lanterns on the beach, right?"

"Yup, that's why they called 'em mooncussers. Couldn't do their work when the moon was out."

"Andrew Harding also ran a small country store at the end of the road. Wrote a book about it, chronicled the comings and goings of a couple of old geezers he called 'good Walter' and 'wicked Walter.' They were Eldredges, cousins I think. Anyway, I guess 'good Walter' paid his bills and 'wicked Walter' didn't."

"So that's where the name Andrew Harding's Lane comes from. I've seen it on the street sign."

"That's probably where the erosion will occur this time too," added Giese.

"Well, I guess the village and old lighthouses are still over there under several fathoms of water. I sure hope it doesn't happen again."

After a few more hours of collecting data, the men clambered into the harbormaster's boat. Hiscock started the engine and picked his way gingerly over the backs of rolling waves while Giese attempted to dry his notes. It was enough to dampen his Southern charm.

"By the way, you ready for next week's meeting?" asked Miller.

"Oh God, *that*. You know this rain isn't much fun but that meeting's going to be pure hell."

CHAPTER 5

An Opportunity Lost
April 27, 1987

Is this the breach? We're not sure. One thing we
do know, the conditions we see associated with
this breach and the dynamics are such that if this
isn't the one, one will be with us very soon.

—*Graham Giese,*
April 27, 1987

An expectant buzz of excitement filled the Chatham High
School auditorium. Two hundred people jammed the aisles
and spilled into the corridors. Most had arrived early to get a good
seat. Hopefully, this meeting, sponsored by the Friends of
Chatham Waterways, would reduce the level of confusion that had
been rising almost as fast as the level of water in Pleasant Bay.

When the breach first opened, most people thought that it
would fill back in and Pleasant Bay would return to the way it had
always been. Within days, however, it became clear that the breach
was not going to close. Every week the breach became deeper and
wider, allowing more water to flow in and out of Pleasant Bay. In a
month's time, the breach was almost a quarter of a mile wide, with
a treacherous, twenty-foot-deep channel that zig-zagged through
it unpredictably.

In early February, many of the people in this auditorium had watched as two local fishing boats tried to pick their way through the narrow cut. The first boat maneuvered too far to the north and had to circle back before running aground. The second boat had lain offshore watching the first, then dashed through safely to the cheers of the hundreds of onlookers.

However, as more fishermen started to use the new inlet, accidents cropped up precipitously. Several hulls were cracked as six-foot waves slammed fishing boats down on sandbars and threatened to swamp them before their owners could maneuver them back into the channel. At the meeting, grim-faced Coast Guardsmen handed out an ominous notice to mariners outlining the dangers that less-experienced boaters would face during the upcoming summer.

For the past three months, the homeowners in the audience had watched as the Atlantic Ocean tore away their land. By April, ocean waves surging through the inlet had eroded as much as 50 feet off the fragile sand dunes that protected many people's property. The Chatham Conservation Commission, which enforces the state's wetlands protection law, had taken a hard line. The commission would not allow homeowners to use seawalls to protect their houses. Already the first of many injunctions against the Conservation Commission had been handed down by the state court. These were difficult decisions. The Conservation Commission was made up of local volunteers, many of them neighbors of people they had to force out of their homes.

On a lighter note, the board of selectmen had banned sun bathing au naturel after a nationwide nudists newsletter had suggested that the now-isolated South Beach would make an ideal new nude beach.

It was against this background that the Friends of Chatham Waterways invited Dr. Graham Giese to address a group of concerned citizens. At the time, there was no better person in the country to give the audience an accurate description of what to expect in the coming twenty years.

Graham paced back and forth before the meeting. He was nervous. By training and temperament he was cautious, perhaps ill-suited to address this anxious crowd. Two hundred people jammed the aisles and spilled into the corridors. They expected answers to their immediate concerns. Would homeowners lose their homes? Would fishermen lose their jobs? An expectant buzz of worry and excitement filled the high school auditorium.

Dr. Giese would have been far more comfortable discussing the vague certainties of what would happen in a hundred years than he was in speculating about what might happen in the next few months. Nevertheless, his talk was illuminating. He recapitulated the history of the 140-year Nauset cycle. He explained that popular schemes to fill in the rapidly growing inlet were hopeless if not nonsensical. They would only cause the inlet to open up further north. The time had come for a new inlet.

However, Dr. Giese failed to detail the implications of his report on the Nauset cycle. The report clearly showed that the new inlet would migrate south and that the locus of erosion would migrate with it. Eventually, erosion would threaten 200 homes and the Chatham Lighthouse. Sand from the inlet would push north, filling in the Chatham fish pier. At least 200 homeowners stood to lose their homes, and 100 fishermen could lose their jobs, their savings, and their investments in boats and fishing gear.

The meeting seemed like the ideal forum to inform people about the problems they would have to face in the near and not-so-distant future. As one frustrated homeowner described it later, "I was shocked. The whole thing was a lost opportunity. It reminded me of the meeting in 'Jaws' where the Chief of Police is pressured by the selectmen to assure the town that there was no reason to panic: 'There are no sharks on Amity Island. The beaches will stay open for the Fourth of July weekend.' It was clear from Giese's report that a lot of us would soon be facing the loss of our homes, but he never came out and said it. He never offered us any advice."

A few weeks after the meeting, I asked Dr. Giese if he felt he should have taken a more active role in explaining the implications of his research. "No, I don't feel as a scientist that I should speculate or offer advice. That's not our role. A scientist's only obligation is to do good research. The rest is up to others."

But other people felt that if Dr. Giese had been more forthcoming he could have helped prevent the paralysis that was to cripple the town for the next five years. It is a common problem. Many people are confused about the role scientists should play in addressing environmental issues. Should they remain in their ivory-towered labs or plunge into the hurly-burly of the political world?

Some people feel that scientists are privileged to receive public money to conduct their research and therefore should feel obligated to share their results and concerns with the affected public. They point out that scientists are often the only people who have studied the problem in enough detail to speak with real authority. On the other hand, scientists are often criticized, particularly by their peers, and sometimes by their funding agencies, if they become too involved in public matters.

But perhaps it didn't matter. Dave Aubrey, Graham Giese's colleague at the Woods Hole Oceanographic Institution would later argue that, "It probably didn't make any difference what was said in 1987. It might just take 5 years for a community to become educated and sensitized enough to deal with a problem like coastal erosion."

So, by the end of the meeting, most people went home feeling happy, relieved and reassured. A few went away feeling frustrated and powerless. However, as the months progressed, no calamities occurred and the skeptics started to doubt their concerns. Had they overreacted? Only time would reveal the answer.

CHAPTER 6

Pleasant Bay

May 1987

> I have noticed in my life that all men have a liking
> for some special animal, tree, plant or spot of
> earth. If men would pay more attention to these
> preferences they might have dreams that would
> purify their lives. Let a man decide upon his
> favorite [spot] and make a study of it.
>
> —*Frances Densmore*
> Teton Sioux Music, *1918*

It was mid-May. The early morning sun streamed through the window, suffusing my room with a soft yellow glow. I lay in bed savoring the quiet music of songbirds. The ancestors of some of these birds had woken me when I was a child years and years ago.

I had lived in this house almost every summer since birth. It had seen me grow and triumph and stumble and fall. It was where my family returned to celebrate the birth of a new child or mourn the death of a loved one. It had given us cohesion and reassurance for almost 50 years.

However, now, this house itself was in danger. Though the new inlet was ten miles distant its presence was already being felt. Each high tide, now six inches higher than before, would remove a few more grains of sand from the bank that supported our house. I had picked this day to set out some stakes at the toe of the bank. But first I would check on the marsh.

I walked down to the marsh, where tiny green shoots of new grass were already sprouting through the stubble of last year's growth. A verdant fringe of new life lined the creeks and marshes of the bay.

At the edge of the marsh there was some dramatic evidence of the effects of the new tidal regime. Hundreds of cedar trees, many of them 40 or 50 years old, were withered and brown. They were being killed slowly by salt water as it bathed their shallow roots during the higher tides. Cedar trees had always symbolized the kind of rugged species that could survive the sand and salt of Cape Cod's barren, often hostile environment. However, the subtle new change in tidal range had revealed how precarious, in fact, was their existence at the very edge of the estuary.

However, I was more concerned about the marshes themselves. Many scientists had predicted that if the sea level rises more than half an inch for three years, marshes would start to die back. If that happened, the thousands of miles of marshes and estuaries that fringe the United States would be killed.

These estuaries are America's forgotten breadbasket. They are ignored largely because they are one of the world's most energy-efficient sources of food. They require neither huge investments in fuel, nor machinery nor fertilizer. Their bounty of fish and shellfish is essentially "free," requiring only a small boat and a strong back to harvest.

Even without human intervention, estuaries are ten times more productive than the most productive wheat field. They provide an annual average of 125 pounds of commercial seafood per acre and support two-thirds of all the foodfish caught in the United States. It would be a huge tragedy if this source of "free" food were to disappear because of sea level rise.

Part of the reason I was looking for evidence that the marsh was dying back was because it would be such a significant discovery. As luck would have it, I found an area right beside our house. It was a small marsh at the end of a creek. Although it was already well into the growing season there was little evidence that the marsh grass was coming back. As I put down stakes to mark the edge of marsh growth, I was already anticipating the stir that this finding would cause.

One of the reasons I was so excited about the changes occurring in Pleasant Bay was because it presented such an ideal environmental experiment. The bay was particularly well known. Students had collected information about the bay for almost twenty years. Now they could use that baseline data to compare the conditions in the bay before and after the new inlet. However, it would take several growing seasons before we could really see the changes in the marsh.

The changes below our house were equally subtle and disturbing. The high tides had eaten six inches into the toe of the bank. If the erosion stopped, there would be no great damage. However, I had seen the destruction caused by the erosion in Chatham and had the disquieting feeling that the erosion below our house would not stop at six inches. I drove some stakes into the beach to mark the toe of the bank, figuring they might help monitor the erosion in the future.

By then it was three o'clock, time to head out onto the bay to catch some flounder. It was there that the changes were most perceptible but hardest to measure. The water had a clarity and freshness it lacked in former years. The tidal currents were stronger as they ebbed and flowed throughout the bay. The bay even looked bigger and seemed to act more robustly. At high tide, the bay covered areas it has never covered before; at low tide, it uncovered sandflats that had never before been dry.

Below the boat, rich beds of new spring eelgrass waved in the outgoing tide. I plucked off some blades of eelgrass and noticed the tiny yellow flowers that were just coming into bloom. Suddenly I realized that I held the key to measuring one of the most important changes brought about by the new inlet.

Eelgrass is an intriguing plant with a curious evolutionary history. Unlike most aquatic plants that are algae, eelgrass is an angiosperm, a more complex plant similar to those found in a home garden. Its ancestors used to live on land, but over time it re-evolved to live in shallow salt water estuaries. However, like other angiosperms it still propagates by releasing pollen and bears these fascinating underwater flowers.

It was because of this curious evolutionary twist that eelgrass had become an important indicator species that can be used to measure water quality. Like other angiosperms, eelgrass needs a lot of sunlight to thrive. If nutrients from septic tanks or road runoff cause algae blooms, water clarity is reduced and eelgrass dies back.

Fortunately, I had just completed a series of maps made from aerial photographs of eelgrass beds taken in 1982, before the inlet opened. They showed that several areas of Pleasant Bay had lost acres of eelgrass because of poor water circulation and nutrient enrichment from too many septic tanks and too much lawn fertilizer.

Perhaps the new inlet would provide Pleasant Bay with the figurative and literal flush it needed to cleanse itself. With rising excitement, I realized that if I could convince someone to take new aerial photos of eelgrass in Pleasant Bay we would have an excellent way to quantify the improvement in water quality caused by the inlet. Dave Aubrey at the Woods Hole Oceanographic Institution had taken the earlier photographs and knew how to get funding. I decided to call him as soon as I returned to shore.

But soon my attention was drawn to another minor drama. As the day drew to a close, a sailboat glided silently down Meeting House River. Suddenly it stopped, its keel firmly ensconced in the soft ooze that covers the bottom of the channel. The boat had to stay for several more hours until the tide turned.

Never before would this boat have run aground in the middle of the channel. From now on, people would have to pay closer attention to the tides. If not, they too would end up spending a few extra hours aground on a sandbar awaiting the turn of the tide—not an altogether unpleasant experience on a beautiful day in early spring.

CHAPTER 7

Summer 1987

Never doubt the power of a small group of
committed individuals to change the world.
Indeed it is the only thing that ever has.

—*Margaret Mead*

Dave Aubrey burst through the metal door of the Coastal Research Center of the Woods Hole Oceanographic Institution. He was late again. He jumped into his jeep, installed his cellular phone and called Scotland.

"Hi, Bob, how did your data come out? Yeah? . . . good. Say,
do you think you can ship the thermistors to Istanbul by next
week? We're goin' to need 'em for our Black Sea cruise."

It was 7:00 A.M., the time Dave usually reserved for returning
his phone calls. It was an annoying habit but one that his respon-
dents got used to. It was the only way to talk to the 44-year-old
director of the center, who managed to coordinate half-a-dozen
research projects strewn across the globe before most people finished their morning coffee.

Today Dr. Aubrey was driving to Otis Air Base to catch a Coast
Guard helicopter. He would fly over the Chatham inlet with Congressman Gerry Studds and several officials from the Army Corps
of Engineers. He liked this kind of work. If everything went well,
he would land an $80,000 grant from the state and part of a

$300,000 contract from the Corps of Engineers to study the dynamics of the new inlet. Hopefully, data from the studies would allow him to prepare a model that could be used to predict the behavior of other inlets in different parts of the world.

Since January, Dave had been commuting between Woods Hole and the University of Virginia. He had been on sabbatical at the University of Virginia where he was co-authoring a book about sea level rise. During his time at UVA he had been playing tag team with Graham Giese, his colleague on the Chatham project. Now Aubrey was back in Woods Hole and eager to plunge into the practical problems of how to deal with the new break.

In Chatham, Cecilie Brown walked slowly to the beach with her two daughters. Fishing had been good the past few seasons. Nick could afford to buy them a new house closer to the ocean.

Cecilie had dreamed of being able to spend long days at the beach while Nick was away at sea. However, it was not to be. By June most of the beach at the end of Andrew Harding's Lane had washed away. She laid the kids' towels beside the parking lot and stared out at the new inlet.

Andrew Harding's Lane had always been a sociable place. True, Good Walter, Wicked Walter and Andrew had long since departed, but they had been replaced by the Wilsons, Rolfes and Galantis. It was a friendly group that often gathered on the beach to share a beer and watch the ocean beyond the inlet. But by the summer of 1987 they were worried.

It was clear that their houses would not make it through very many more winters. Now their beach-side discussions turned to talk of politics, lawsuits, and agitation. They huddled on the remains of a beach in front of Steve Rolfe's cottage. It was the first summer home built in Chatham.

Steve Rolfe's entire life was wrapped up in this cottage. His parents had sent him through college on the summer rents they received from the house. "It's all I got, all my savings are in this house. I don't have much of a job but I love the ocean. I'll do anything to stay here."

Joan Wilson, who had lived in the adjoining house all her life, said, "Peter Mason wants to start a citizens action group. Has a lawyer in mind, sharp son of a bitch. Think you'll join?"

"Yeah, I think I will. Albert Nelson stopped me on the beach last week. He said he'd join and bring in thirty more families from south of the lighthouse," added a neighbor. A minor insurrection was in the works.

But mostly the summer of 1987 was a time to wait, watch and plan. September would bring the first true test of Chatham's resolve.

CHAPTER 8

Andrew Harding's Lane

September 28, 1987

Malcolm coughed and stared into the distance.
"Let's be clear. The planet is not in jeopardy. We
are in jeopardy. We haven't got the power to
destroy the plant—or to save it. But we might
have the power to save ourselves."

—*Michael Crichton,*
Jurassic Park

It was a brilliant, clear, blue autumn afternoon. Sunlight flashed
off the crests of far-off waves and puffy white cumulus clouds
floated through a cobalt sky. Cape Cod was in the grip of its most
beautiful season.

There was no wind, no storm, no surf, only small wavelets lap-
ping gently on the Chatham shore. I had come down to Andrew
Harding's Lane to enjoy the day, but gradually it dawned on me
that something was wrong; something powerful, strange and irre-
sistible was happening.

Instead of receding, the tide kept coming. As it rose each small
wavelet scoured out an inch or two of sand from beneath the asphalt parking lot. Gradually a four-foot piece of asphalt sagged
then collapsed and the waves started undermining a new area. The

process continued as the tide rose inexorably higher. Each new incursion raised a disturbing question: Would it ever end?

Then, as subtly as it had begun, the process slowed. Each wave slid a shorter distance up the beach. The tide had finally turned, but it had done its damage; in less than 30 minutes, 15 feet of the parking lot had been destroyed.

It was certainly not the most dramatic day I had ever experienced on this coast. I had watched at least a dozen hurricanes and major storms slam into this fragile peninsula. But there was a certain logic to a vicious storm. It made sense to see destruction when the wind was lashing your face, when waves roared as they dragged boulders out of exposed cliffs.

However, to watch the ocean casually, almost tenderly, destroy a parking lot on a calm, warm, sunfilled day was a profoundly unsettling experience. The ocean had not encroached on this shore for 140 years and now it was destroying it with the same disdain, the same casual nonchalance we might have ascribed to it thousands of years ago when it helped build this coastal dune.

In years to come, as the coastal bank in front of my family's house started to erode, I would become acutely aware of the subtlety of the process. I would lie in bed at night, fitfully watching the full moon, as hour by hour it rose to eclipse Venus and Jupiter. I knew that the three celestial bodies were unknowingly, uncaringly pulling our planet's oceans out from one side of the world while equally unconsciously the sun pulled the oceans out from the other side of the world.

In an environment that constantly reinforces our self-image that we are the species that controls the world and its future, it was unsettling to have these faint reminders that circling about in space millions of miles away are massive bodies that are engaged in an unintentional conspiracy to shape our lives. Just this chance alignment of the planets would have a profound impact on my life. It would raise the tides to a 60-year high, eroding half an acre of our land and jeopardizing our home.

Such are the primordial fears raised by coastal erosion. Perhaps more than any other environmental concern, the spectre of the sea rising uncontrollably exposes our conceit that humans are in control and can conquer nature. It even calls into question our equally hubristic belief that we alone are responsible for the world's woes and that we alone can solve them. In reality, when it comes to our influence on the biosphere, humans are like physicians who try to heal an ailing patient. We can run some tests, diagnose a disease, perhaps prescribe some antibiotics or perform surgery. But ultimately, like the human body, it is the biosphere that must heal itself, or not. And ultimately, of course, it may be that we are the planet's problem, the pathogen that the world must shake off or subdue, before it can heal itself once again.

CHAPTER 9

The End of the Beginning

December 10, 1987

The evidence now suggests that the beach house
may go the way of the buggy whip.

—Michael Robbins,
Oceans *magazine, 1987*

It was a cold day in early December. Jeff Benoit pulled up the
sleeve of his parka. Around the Coastal Zone Management office the director was known more for his bow ties than for his worn
parka. He scooped a handful of sand from the cliff and handed it to
Liz Koulheras. "Looks like dune sand too. I'll have 'em check it
under a scope to be sure."

Jeff Benoit and Liz Koulheras were in Chatham to meet with
the Andrew Harding's Lane homeowners. The homeowners had
hired an engineer and a lawyer and formed a citizen's action committee called BREACH to deal with state and town officials.

The procedures had been complicated. Before the homeowners could build or remove anything within 100 feet of a wet-
land, they had to file a notice of intent with the local Conservation
Commission. Their request for a seawall had been turned down, so
they were entitled to this on-site meeting with state officials from
the Department of Environmental Protection and the Coastal
Zone Management Office, who would decide their fate.

Liz Koulheras was head of the wetlands office of DEP and Jeff Benoit was head of Coastal Zone Management. He was a suave career civil servant who had received his M.A. in coastal geology from the Skidaway Institute of Oceanography in Georgia.

In coastal geology circles, Skidaway was famous for being the birthplace of the school of coastal retreat. In 1985 Orrin Pilkey of Duke University had invited a group of geologists to meet at Skidaway to prepare a national strategy for beach preservation. They draw up the Skidaway Statement, which called for coastal retreat in the face of sea level rise.

Benoit had remained faithful to the Skidaway Statement and was well-versed in the recent literature on sea level rise. He was familiar with the 1980 study that predicted that the sea level would rise 24 feet during the next hundred years. It included maps showing New York and Washington inundated by the oceans. But Benoit went by the more conservative EPA report released in 1986. Its worst-case scenario predicted that the sea level would rise 12 feet during the next hundred years.

He was also familiar with popular coverage. During the summer, *Time* magazine had devoted a cover story to coastal erosion. *The New York Times* followed suit, with an article about sea level rise that featured the Chatham break. The articles made Liz Koulheras and Jeff Benoit's job that much easier, because they made drastic sea level rise seem like a certainty.

Throughout the afternoon Koulheras and Benoit played good cop, bad cop. While Benoit sympathized with the homeowners' plight, Koulheras told them they would not be allowed to build a seawall to protect their homes. It was a dramatic showdown between the homeowners and the state, but the next act would be staged by mother nature.

By the end of December, one hundred more feet of the beach in front of Andrew Harding's Lane had disappeared. The high tides of winter had undermined two houses and the Atlantic Ocean was licking at the foundation of the Galanti cottage.

PART II

The Middle Years
1988–1991

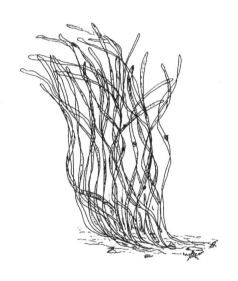

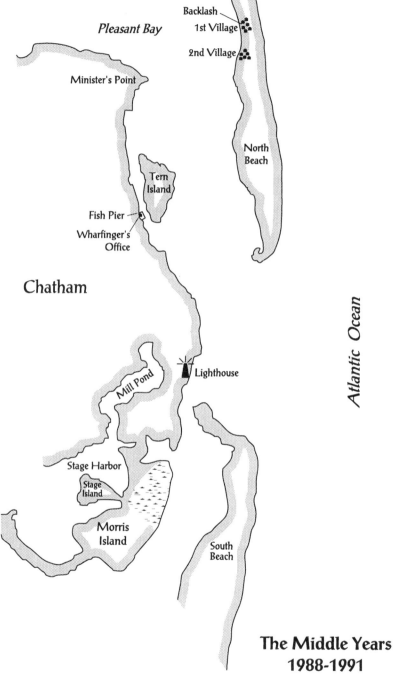

Pleasant Bay

Backlash

1st Village

2nd Village

Minister's Point

North
Beach

Tern
Island

Fish Pier

Wharfinger's
Office

Chatham

Atlantic Ocean

Lighthouse

Mill Pond

Stage Harbor

Stage
Island

Morris
Island

South
Beach

The Middle Years
1988-1991

CHAPTER 10

Nick Soutter

January 15, 1988

Being head of BREACH was like being head of a
Latin American country. You've no idea what's
going on in the outlying districts.

—*John Whelan,*
in Breakthrough
by Tim Wood, 1988

Nick Soutter liked the long drive to Chatham. It gave him a
chance to think and brush up on languages. He reached forward to flip his Russian language tape to the other side. He had
been working on the dative case for the past two weeks. Perhaps it
would help him with an upcoming client, but he doubted it.

"How many Russians are likely to come pounding on my door
for legal advice? Not too damn many. Besides, how much money
do the fuckin' Russkies have these days?"

Soutter unlatched the glove compartment and a cascade of language tapes clattered to the floor. Portuguese, Spanish, French,
German; at least some of these might help his practice. His clients
were usually poor, black and in trouble. Most didn't speak any En-
glish. But Soutter had gained the respect of both his clients and the
prosecutors who opposed him because of his vigorous defenses.

51

"This case should be a piece of cake. I take three grand as a retainer, spend a few weeks helping these poor buggers get their seawall and go home. No sweat, no trouble, no problem."

On the phone Peter Mason had made it seem pretty desperate. Probably just exaggerating. Soutter swerved off Shore Drive. "This must be it. Unpaved driveway, the whole nine yards. The windmill should be up ahead." He got out of his car and climbed the stairs.

"Jesus," thought Soutter as he entered the room. He had never seen such a forlorn little group of people. They looked like war-weary refugees. They were huddled inside the tiny room over-looking the raging Atlantic. Soutter knew the look. It was the look of people who felt totally outmaneuvered by a legal system they couldn't fight or understand.

"They all had shell-shocked expressions, as if they had just been run over by a truck. There was a sense of total hopelessness in the room. I figured the first thing was to try to boost morale. I told 'em they ought to sue their asses. Give 'em a wake-up call. They'll get the point."

John Whelan was the leader of BREACH, but Peter Mason was the most worked up. He wanted to push ahead with a rock wall instead of sandbags, the only temporary solution the state would allow.

Whelan questioned Soutter. "What do you know about environmental law?"

"Not a damn thing. But I know this. They'll pay attention when you hit 'em with a suit for ten million bucks."

"Could we ever win?" asked Steve Rolfe.

"You're not going to get a penny, but you'll have the pleasure of waking 'em up."

"They're goin' to argue a lot of technical points. Do you know about the dune/bank distinction?" asked Whelan.

"What the hell is that?"

"Benoit is saying half the houses are on a dune and the state regulations say you can't build seawalls in front of a dune but you can in front of a bank."

"That's absurd. How do they know the difference?"

"Size of the sand I guess. How would you argue against it?"

"I'll just say there's not enough sand there to tell if it's a dune."

Peter Mason chimed in, "I say we should give him a try. Why not let him go out and talk to Channel 58?"

That was the first inkling Soutter had that he had been hired.

If there was one thing Soutter was good at it was thinking on his feet. The first thing the anchorwoman asked was about the dune/bank distinction.

"I didn't know one damn thing about the dune/bank distinction," Soutter recalled later. "But there is an old saying in trial law, 'When all else fails argue quantity,' so I went on about there not being enough sand left in front of the houses to call it a dune. Two weeks later I looked like a goddamned genius."

CHAPTER 11

The Beginning of the End

January 21, 1988

I still choke up a bit even though it is over four
and a half years since we lost the cottage to the sea.

—*Paul J. Galanti, Letter to the Editor,*
The Cape Codder, *August 1992*

It was January 21st, 1988. Chief of Police Barry Eldredge
stamped his feet and pulled his visor down against the cold rain
that drizzled out of the low gray clouds. One of his officers beck-
oned for him to go to the crowd barrier.

"Chief, this man says he just drove 300 miles from Connecti-
cut and wants to take some pictures from the shore."

"Sorry, sir, tide's comin' in. I can't let anyone on the beach till
it goes back down."

Another helicopter swooped low to film the small crowd gathered on the shore. Journalists in city shoes clambered over the slip-
pery rocks. Eldredge hated having the world snicker at his town as
it tore itself apart and he was damn sure he was not going to be
portrayed as some sort of local yokel, hick cop beamed nationwide
on CNN.

The Chief knew his town, his tides and his boats. For three
generations his family had made a living from shellfishing and boat

building. His great-grandfather, Oliver Eldredge, had started the old boatyard "down ta Stage Harbor."

Eldredge didn't like being put in the middle and it seemed like he had been in the middle of this damn mess since last August. There hadn't been any major storms, but that hadn't mattered. Moderate winds and the regular high course tides of Fall had done their damage. By late September they had eradicated 200 feet of the dunes and beaches that used to protect the first line of shorefront homes.

The press had been alerted that they would soon get some good video, a house would soon be washed away and people were upset with the state.

Everyone in town knew it was just a matter of time. Tension had been mounting between town officials and homeowners who had tried to get permission to protect their homes. Soutter had demanded that his clients be given permission to build seawalls. When they were turned down, he sued the state for ten million dollars. People were getting a taste of Nick Soutter's city slicker style.

Eldredge couldn't blame the homeowners. In his line of work he had seen people lose their homes to flood, fire, wind and storm. It was never easy and this time it was made a hell of a lot worse by the presence of this pack of sensation-seeking journalists vying for flashy video clips at the expense of his town.

The last two weeks had been pure chaos. First a judge slapped the state with an injunction that allowed the homeowners to truck in boulders, then she had reversed herself. The next day Chatham was full of rumors that the homeowners would try to sneak a convoy of trucks into town anyway. Town counsel had ordered the police to stop the trucks, but somehow they sneaked into town at dawn and dumped the boulders on the beach surreptitiously.

Now the shore was swarming with scores of journalists. Antennae from their broadcast vans swayed above the lighthouse parking lot like masts above a harbor. It was as though a sadistic death-watch was taking place.

Eldredge watched as demonstrators passed out literature and

spray painted graffiti about various members of the Conservation Commission on the side of one of the doomed buildings. An anchorwoman tottering on her high heels interviewed Nick Soutter and Steve Rolfe in front of the Galanti cottage. Waves had been undercutting its foundation for several days. Suddenly the house shuddered, creaked and slumped into the water.

"Whoops!"

"Damn!"

"Oh, shit!"

The brick chimney clattered to the beach like so many children's blocks. The Atlantic had taken its first victim in over a hundred years. Other homes were soon to follow.

CHAPTER 12

The Old Rolfe Place

February 1988

I think about it all the time. I guess you could say
I'm still preoccupied with losing our house,
particularly around the holidays.

—*Steve Rolfe,*
November 24, 1992

T he Rolfe's cottage was one of the first summer homes built in
Chatham. No one knew exactly when it was built because a
fire had destroyed the deed at the old county court house in 1900.
It was a small cottage with three bedrooms upstairs and a living
room, den and kitchen downstairs. It was supported by 24 stilts
sitting on flat pieces of Quincy granite that were buried in the sandy
soil. Steve Rolfe knew about those pieces of granite because he and
his wife Debby had replaced the stilts one by one over several summers.

His grandfather had bought the cottage the year of the '38
hurricane. "I can still remember him walking up the hill to the
New Yorker to drink with his old cronies. That was before they
renamed it the Squire."

Steve Rolfe paused to reflect. He had a sturdy, burly build and
a full beard that reminded his wife of Grizzly Adams.

Storms hadn't been good to the Rolfe family. Steve's grandfather had died in another hurricane when his car skidded into a tree in Harwich.

Steve Rolfe had started living in the cottage the year he was born, in 1948. "We never had a TV in that cottage, only had a phone the last few years. If we wanted something to do we had the beach out front, or we could walk down to the fish pier or up to the lighthouse. My father would drive down to the cottage, park the car and hardly drive it all summer. We had everything we needed within walking distance. It was the perfect location.

"A neighbor from up the road had an old station wagon with wood sides, a 'woodie' I guess you could call it, but we never did. We'd drive up to Orleans, buy some bait at the Goose Hummock Shop and drive out to the beach. It wasn't four wheel drive in those days. You had to get out and take the air out of the tires to drive down the beach. We'd spend all day out on North Beach fishing for bass and bluefish.

"When we got older, my parents would rent the house for June, July and two weeks in the beginning of August. Our family would use it for the last two weeks of August and as much as we could in the Spring and Fall. All five of us kids went through college on the rent from that cottage.

"Debby and I spent our honeymoon there. We loved that cottage. We re-shingled it, painted it and put in insulation. As a matter of fact, we really built our life around coming down here. During the winter my wife would teach and I'd bang nails and take care of the kids. In the summer we'd come down here and she'd take care of the kids and I'd install swimming pools. Did it just so we could afford to stay down here.

"We'd come as soon as we could in March. The cottage would still be freezing. You'd have to go outside to warm up. Art Gould used to run his ferry service out of Andrew Harding's Lane. We let him keep his boats under the cottage so he'd let us borrow one whenever we wanted.

"We'd row out with a box of sandworms and start to catch so

many flounder we'd throw back the small ones and keep the big ones. By the end of the day we'd have filled up several bushel baskets full. We'd row back, stow the boat back under the house and start to fillet the fish. We built a small cleaning table beside the cottage right next to the outside hose.

"You never quite forget how to fillet a flounder. It's like learning to ride a bike; once you learn it you'll never forget it.

"After filleting all the fish, we'd pour oil in a frying pan and heat up the stove. We'd dip the fillets into cornmeal and drop 'em on the hot skillet. Fillets would be so fresh they'd curl up at the edges as soon as they'd hit the hot oil. Pretty soon the whole cottage would fill up with the smell of fish, butter, steamed clams and beer. We'd eat so much we couldn't move, then wander off to bed, and still have enough to freeze for later . . ."

"I didn't know how bad the situation was with the cottage until we saw it on the news. I was there the day the Galanti house went in. I was trying to get everything out of the cottage before it went in, too. I remember one woman coming up with tears in her eyes, saying her husband had proposed to her beside our cottage. If the walls could only talk, there'd be a lot more stories that cottage could tell.

"The day after Galanti's house went in it was calm and we were saved. We started puttin' the cottage up on timbers to move it back behind Effie Butler's place. This might sound funny, but it was probably the saddest day of my life. I still think about it and the damage it has done to our family. My father started to get really depressed and to drink then. He vowed he would never go back to the Cape again." Steve Rolfe paused to sigh.

"Less than a year later we had to destroy it. You know, that simple little cottage used to be like a magnet drawing our family together. Now it's gone and I haven't seen my sisters for years."

CHAPTER 13

Summer 1988

I'm 99 percent certain that we are experiencing global warming.

—*Dr. James Hansen,*
Goddard Institute for Space Studies,
Senate Hearing on global change, 1988

The summer of 1988 was the hottest summer on record. It was the summer that corn fields parched in the Midwest and the Mississippi almost ran dry.

It was the summer that an old farmer cried on national television as he sifted powdery soil through his dry gnarled hands. It was the summer that barges ran fast aground in the Mississippi River.

1988 was the summer during which a climatologist testified to Congress that he was 99 percent certain that the century of global warming was at hand. It was the summer that Al Gore ran his first presidential campaign on environmental issues. It was the summer that fear about global warming reached its peak.

National policies impinged on Chatham events in 1988. A month after Paul Galanti lost his house, President Reagan signed Upton-Jones, a bill that would have paid Galanti 110 percent of the value of his house if he had destroyed it before it had washed away or 40 percent of the value of his home if he had moved it first. "No wonder I've been a Democrat all my life," observed Galanti.

But by the end of the summer people were beginning to wonder if it wasn't more merciful to lose your house all at once in a storm rather than experience slow death at the hands of bureaucrats.

In early summer an official from the Federal Emergency Management Administration came to Chatham to tell Steve Rolfe he could move his house. After he had moved the building twice, the state Department of Environmental Protection ruled that he couldn't build a seawall in front of an empty lot. A few months later the town ordered him to destroy his home.

"They take control away from you. Everyone gave us a different story. The Upton-Jones bill was so new the FEMA guys were making up the rules as they went along. They led us to believe we could move the house back, then changed their story and we lost our house. I really felt we got screwed."

In May the state held a "summit meeting" with the appropriate town, state and federal officials to decide Chatham's fate. In what the *Chatham Chronicle* described as "an unremarkable meeting," the officials decided to work together to expedite harbor dredging and put together a long-range plan the town could refer to over the next 50 years as North Beach and the harbor continued to change.

Dave Aubrey was disappointed that it would take six years for anything to come from the meeting. "I was sick of seeing the lawyers get all the money. If we had started the process then a lot of the later confusion could have been avoided."

Nick Soutter spent most of June and July in the Barker Library at the Massachusetts Institute of Technology studying up on "everything I had been arguing about but didn't know anything about in coastal geology."

Fishing was good for Nick Brown during the summer of 1988, but like everyone else he was hampered by the rapidly shoaling channel.

CHAPTER 14

Wits and Water

November 20, 1988

That fish was somewhere between a dinosaur and
a nightmare.

—Nick Brown,
November 20, 1988

The alarm jangled in Nick Brown's ear. He groped for the
clock; 2:00 A.M., still dark, "Time to make the donuts," Nick
muttered. Cecilie breathed lightly beside him. In the old days she
would have gotten up to share a few cups of coffee. But now, with
the kids and the new house, she needed all the sleep she could get.

What Nick really wanted to do was crawl back under the covers
and reinforce the love they had made the night before. How many
times had they reenacted the wonderful ritual? Never once had he
regretted his decision to marry Cecilie. Nor had he ever regretted
his decision to take up fishing after graduating from business school.

He walked slowly to the window and gazed at the stark out-
lines of cedar trees barely silhouetted by the half moon. Dawn
should break clean and calm in the wake of last night's storm.

"Should be able to make it through the new cut, wind's shifted
to the West," he thought to himself.

After gulping down two cups of coffee, Brown loaded the

pickup. He loved driving through the sleeping village. By day it belonged to tourists and retirees; by dawn it was still the domain of working people.

One by one the pickups of other fishermen converged on the pier. They gathered around Nickerson's Fish and Lobster sipping coffee, joking, loading their skiffs and preparing for the day.

They were the rugged individualists of this seaside community. The town's mythology rested on their shoulders. They shared the early-morning camaraderie, the bond of knowing that they were out doing men's work while tradesmen were still dreaming the dreams of respectable people.

Soon they would be out on the fishing grounds. The camaraderie would be maintained in the conversations on their radios but the competition would be in earnest. The captains of each boat had spent perilous days probing for the best location to set their nets and they were not about to give up their secrets lightly.

Brown gathered up his belongings and packed them in the stern of his small pram. He was careful not to let anyone see the shotgun hidden in a piece of old canvas. It was still early. He rowed to the back of Tern Island and carefully unwrapped the gun. Slowly he advanced from thicket to thicket. Suddenly he crouched, then rose to fire. Two ducks exploded from a small salt pond and flew overhead; the last one fell.

The echo of the blast ricocheted around the harbor. Fishermen looked at each other and smiled, "Brown got another one, right under our noses."

"That was a nice way to start the day," Brown grinned as he started up the *Synergistic* and idled into the pier to pick up his crew.

Dick Bartke stowed the gear while Dana Nickerson stood bow watch. They entered the inlet, then jogged south into the teeth of the Atlantic Ocean. The recent storm hadn't seemed to alter the position of the channel but it was shoaling every day. In rough weather it could be damn tricky.

There are two styles of fishing out of Chatham Harbor. Most of the old guys still liked to go for the big trip. They were full of "piss and vinegar," went out farther and worked the hard tides. To Nick it was a "burn out scene." He preferred to stay closer inshore and take a week off to go hunting during the full moon when the tides were running hard.

The weather held as they steamed out to Crab Ledge. They had left the gill net overnight to fish through two slack tides; that was when the nets caught most of the fish. On Crab Ledge they retrieved the net, which held 5 boxes of cod and pollack.

"Hey, Nick, remember last spring when we got that sturgeon? That was some ugly fish."

"It was strange looking all right, cross between a dinosaur and a nightmare." Even though it was unprotected and over six feet long Nick didn't save it. "I just couldn't keep something that strange. Kinda liked him, to tell the truth."

After Crab Ledge, they steamed to The Lemons, The Mussels and Great Hill. Great Hill was the last fishing ground where you could still see Chatham. Actually, you couldn't see the land—only the Great Hill water tower that loomed over the high school. That was usually as far offshore as Nick ventured.

By three o'clock the nets were reset and Nick headed the *Synergistic* back to Chatham. The wind was starting to blow out of the East. "You think we should take the old Chatham bars inlet? It's gettin' pretty sloppy."

"Not enough time; the truck to New York leaves at 8:00 sharp. We'll have to chance takin' the new inlet."

Dick and Dana stood on the bow looking for the channel while Nick studied the inlet. Waves were breaking over an underwater sandbar created by the recent storm. Nick steamed south then jogged quickly north. The submerged sandbar broke up the waves and they found a deep trough behind the bar.

Dana finally spotted the darker water of the main channel. "Go for it!"

Nick gunned the *Synergistic* into the calmer waters of the harbor. They had made it through the new inlet once again. Ahead of them a dredge sucked sand out of the spar channel.

"That's a welcome sight."

"You have to hand it to the selectmen. They worked hard to get town money for dredging."

"Damn retirees control town meeting. Cheap bastards don't pay a nickel unless you push 'em."

"It won't last long, though. If the Army doesn't take over, this harbor will be gone in five years."

The Army Corps of Engineers had spent almost two years coordinating and preparing a general investigative study of the new inlet. If the cost-benefit analysis turned out positive, the federal government would build shore revetments and take over maintenance of the channel.

"The Feds have already spent $150,000 on the study. I just wish they'd 'ave spent it on dredgin'."

"Doesn't work that way. If the first study works out, they'll have to spend twice that much on the next one."

"Hey, Nick, you know about this stuff. You goin' to get involved?"

"Nah, got to keep this boat runnin' so you guys can have a job, right?"

"If the Corps turns us down, none of us will have a job."

"I'm just a fisherman now. Besides, I forgot most of the accounting I learned in college. Come on, let's get those lines. We've got a boat to unload."

CHAPTER 15

Hoisted on My Own Petard

February 10, 1991

> My earliest lessons on environmental protection
> were about soil erosion on our family farm, and I
> still remember clearly how important it is to stop
> up the smallest gully "before it gets started."
>
> —*Al Gore,*
> Earth in the Balance

February 10th, 1991, was a bitterly cold day. A north wind swayed the scraggy pines and the sun stayed hidden behind low gray clouds. I was standing in our driveway making stilted small talk with several members of the Orleans Conservation Commission. We were waiting for the perennially late state officials. Suddenly it had become my turn to run the gauntlet.

I had watched as, house by house, the homes of the Wilsons, the Rolfes, the Perrys, the Butlers and the Coxheads had been destroyed or condemned on the basis of the dune/bank distinction. I had also watched as the Chatham Conservation Commission and the state had gradually agreed on revetment designs and allowed other Chatham homeowners to build seawalls to protect their homes. It had been an interesting, though academic, experience. I had the unsettling feeling that some people had paid an extraordinarily high price to maintain a well-meaning but ill-conceived environmental policy. But I was about to learn just how wrenching the process could be.

Finally, I heard the bumps and scrapes of a car lurching down the rutted road. It rolled up a slight incline at the end of our driveway and Jim O'Connell from the Massachusetts Coastal Zone Management Office clambered out.

"Morning, Jim," I said merrily. "You got egg all over your face. What'd you have for breakfast?" Nobody seemed to appreciate the wisecrack.

O'Connell had visited our house before. In October my family had applied to the Orleans Conservation Commission to install a gabions, a low wall of rockfilled wire baskets to shore up the toe of our eroding bank. Jim had told us not to worry, the erosion problem was over; at the most we would lose only 2 feet off the top of our bank. We were denied. Two months later another storm struck. We lost half an acre of land, several dozen trees and 12 feet off the top of our severely eroded bank. Our house was now only 52 feet from edge of the bluff.

Several tons of sand and clay had cascaded down the bank into Pleasant Bay, smothering valuable shellfish beds and the fringing marsh. Though nervous, I was fairly sure we were going to win this round. After all, these were conservation officials and they should be worried about the loss of highlands as well as wetlands.

To understand the problem of coastal erosion, it is necessary to understand the place that banks and bluffs have always held in the psyche of Cape Codders. For generations Cape Codders have been brought up to value and protect highlands. Anyone fortunate enough to own land on a bank takes special care to see that it is well protected. They have a visceral understanding that these fragile remains of the last glacier are their last defense against the ocean.

When I saw the first six inches of erosion in our bank, I panicked. Cape Codders are taught to react to erosion in the same way that the little Dutch boy who held his finger in the dike did.

Growing up on the Cape, my family had few rules. You could get away with a lot. However, the one rule that was sacrosanct was, "Never walk on the bank." In fact, the sanctity of the bank was so well ingrained in me that it often emerged in my nightmares. I would dream of huge rogue waves tearing at our bank while dirigibles and fighter planes strafed us from above.

My sister had a variation on this dream. In her dream she sees our house surrounded by swirling water, but she observes it from above. She flies overhead in what we used to call an octopus, our favorite ride at the visiting summer fair. She swoops and spins around the house, seeing it from all angles and views. In one such dream she saw in an instant all the ways she wanted to remodel our house. I don't know what I fear most: the loss of our bank or my sister's remodeling.

As the members of the conservation commission tramped down the face of the bank, I joked that no one in our family had ever been allowed to do that. Nobody seemed to appreciate that wisecrack, either.

The evidence of erosion was plain to see. The stakes I had put in at the toe of the bank four years ago were sticking out of an empty expanse of sandy beach. The toe of the bank had eroded over 8 feet. Uprooted trees lay crisscrossed on the beach and six inches of sand spread out over the marsh and clam flats.

The state had to determine whether the bank was a significant source of sediment that would build up downstream beaches and whether our house was in imminent danger. Only if the bank was not a significant source of sediment and if the house was in danger could we get permission to slow erosion. We wanted permission to install several rows of the wire baskets that would slow down the erosion and allow us to live in the house for another 60 years. We felt that such a structure would not harm downstream beaches. In fact it would help save more marshes and clamflats from being smothered.

In Massachusetts, erosion control is regulated by the Wetlands Protection Act. The act was cobbled together in 1978 from two separate pieces of legislation, the Jones Act covering salt water habitats and the Hatch Act covering fresh water ones. So today a single act applies to habitats as disparate as a fresh water lake, a rocky coast, a barren barrier beach or a productive estuary.

One problem with this approach is that fresh water lakes, rocky coasts and sandy barrier beaches are relatively barren, whereas an estuary is ten times more productive than the best wheat field. This puts conservation commissions in an uncomfortable position. They are required to give priority to the role of sand as a source of sediment to downstream beaches. Sometimes this means they have to sacrifice a bank that protects a house and cause shellfish and marshes to be smothered, all for the sake of the often hypothetical role the sand will play in building up a future beach. This is a little like prohibiting a midwestern farmer from slowing erosion on his wheat fields so that his topsoil can flow down the Mississippi to help create new oil fields in the Gulf of Mexico in the next millenium.

The reason that the Wetlands Act only regards erosion as a positive force is that it was designed by geologists. Geologists get very excited when they see change because they see it so seldomly. Unfortunately, humans and most of the things we care about live in biological time, not geological time. While a hundred years might be but an eyeblink to a geologist, it is crucial to a homeowner who is deciding whether to save his house, a shellfisherman earning his living, or a town official planning for future tax revenue.

We were turned down by the state in March. That night was one of the bleakest in memory. I was flooded with feelings of remorse. My entire family had put their faith in me and I had let them down. It was a tragedy that would reverberate through our family for generations. I could already hear the creation of family myths: "Oh, yes, wasn't he the one who lost the house on the Cape? Always was a bit of a rotter, that one."

It was also a time to reflect on my previous attitude. When the inlet first broke through, my immediate reaction had been, "Oh great, now we can get rid of those houses and return some open land to the public." In my initial inclination to think of the public good, I forgot to think of the individual harm. Now that cavalier attitude had come back to haunt me.

But I had one more source of regret. In my eagerness to discover exactly what was happening because of the new inlet, I had driven the stakes in the beach along the toe of our bank.

Usually when the conservation commission makes a site visit, everyone looks around, shrugs and says, "Looks pretty bad, don't it? Guess we should let 'em build a seawall." Unfortunately, in our case, everyone had seen the stakes that showed only six inches of erosion per year. The state could dismiss the storm that caused 12 feet of erosion as "an anomalous episode that was unlikely to recur in the next 60 years." They based their decision on the long-term rate of erosion that was revealed by my stakes. I had been hoisted on my own petard.

CHAPTER 16

The Chatham Fishermen's Wives Coalition

Summer 1991

It is becoming apparent that there are no economically viable solutions to the commercial navigation problems in Chatham Harbor.

—*Colonel Philip R. Harris,*
Army Corps of Engineers,
Letter to Congressman Gerry Studds,
July 29, 1991

Su-Ann Armstrong was fixing breakfast when the first shoe dropped. On the front page of the *Cape Codder* was a picture of Chatham selectman Andrew Young, Colonel Harris of the Army Corps of Engineers and Congressman Gerry Studds. The headline read, "Army Reconsiders Dredge Plan for Chatham Harbor."

She reached for the marine radio beside the sink. She used this every day to talk to her husband when he was at sea. Sometimes he would ask her to buy some supplies or arrange to have the boat's oil changed; other times she would just call to have him say goodnight to the kids.

"Chris, this is Su-Ann, do you read me? Over."

"Morning, Su-Ann, read you loud and clear. What's up? Over."

"Chris, it looks bad. This morning's paper says the Army Corps might not dredge the harbor. Over."

"Jesus! Over."

"Something about the cost-benefit ratio not working out. Over."

"How can they do this to us?"

"You better call the rest of the fleet. See what Nick Brown thinks. I'm going to call Cassie and Shareen to see what they know. Over."

The news was a bombshell. Chatham had put up half of the money for a $220,000 study by the Army Corps of Engineers. The fleet had assumed that it was largely a formality. The Corps had led them to believe that as soon as the study was finished the Corps could step in and take over dredging the harbor. So far Chatham had spent $162,000 of its own money dredging the harbor but it was filling back in at the rate of six inches a month.

During the winter most of the boats had to moor outside the harbor and offload their catch into small skiffs, then row it to the dock and offload it again. It added a lot of dangerous extra time at the end of a long day. In December Jack Our's skiff had tipped over and he had fallen into the icy waters. His heavy waders had pulled him under twice before a mate from another boat was able to haul him out.

The town could only afford $500,000 for dredging and that would fix the problem for only about 5 years. Selectman Young had explained, "That's it. When the money's gone the town will have to abandon the harbor. We will lose $30 million dollars a year and 500 people will be out of work."

A week after the announcement, the selectmen held a meeting with the Army Corps of Engineers. It was held at 10 A.M. on a good day for fishing. The fleet was never notified and no fishermen were in attendance. It was clear someone would have to fill the vacuum and, as is so often the case, it was the fishermen's wives who stepped forward.

Cassie Abreu's kitchen became command headquarters. She came from a fishing family and was Chatham's former shellfish warden. The fishermen's wives manned the phones every day. They collected information, called officials and warned the fleet to attend the important meetings with the Army Corps of Engineers. Gradually the telephoners became the nucleus for the Chatham Fishermen's Wives Coalition.

Shareen Eldredge never liked the sexist connotation of the name but she had to admit that being a fisherman's wife was different. Traditionally, fishermen's wives have had three jobs: wife, single mother and business partner. They were the ones who kept the books, paid the bills, and sought out the best medical and business insurance. Some of them had worked on their husbands' boats, helped bait gear and sold fish. Many of them knew the business better than their husbands did. When it became clear that they would have to attend meetings their husbands couldn't afford to go to, it was seen as only another role they had to assume. But the wives never realized the power they would wield nor the feats they would soon accomplish.

An earlier August meeting between fishermen and the Army Corps of Engineers had been emotional. Shareen Eldredge remembered the fishermen and the Corps talking "apples and oranges." Two beleaguered economists from the Army Corps of Engineers had tried to calm the fishermen, who saw their livelihoods at stake. Some fishermen had just invested $250,000 in new shallow-draft boats that could deal with the conditions in the new inlet.

The economists kept insisting they had to look at the national situation. What they didn't mention was that the Bush administration had been trying to scuttle the Rivers and Small Harbors Act. The administration couldn't get Congress to kill the program, so they had changed the rules. Suddenly the fact that the outer inlet was only 6 feet deep made it impossible to meet the cost-benefit analysis. No one had mentioned the depth of the channel before.

Gradually it dawned on the wives that, if they were going to win the fight in Chatham, they would have to win the fight in

Washington. They would have to overturn national policy. Maybe they would even have to persuade Congress to pass a special bill. They turned to Congressman Gerry Studds for help.

Fishermen had never forgotten what Gerry Studds had accomplished for them. As a freshmen congressman, he had persuaded Congress to adopt a 200-mile fisheries boundary. It had essentially increased the country's territory by one-third. Even when Studds was censured for having an affair with an underaged congressional male page, fishermen had continued to back him.

It was an incredible political feat, but it also reflected fishermen's respect for independence. They didn't care what Studds did with his sexual life as long as he did his work, protected their interests and brought home the bacon.

The second shoe fell on August 19th. Hurricane Bob caused over a billion dollars worth of damage to New England. Thousands of trees were blown down and boats were damaged. However, the hurricane arrived and departed at low tide. Chatham homeowners breathed a sigh of relief, fishermen were glad that they had lost only a few days of fishing and almost everyone was able to cut up enough firewood for the coming winter. The term "windfall" gained new meaning, and the fishermen's wives continued their work.

The real storm was yet to come.

CHAPTER 17

The Halloween Storm: North Beach

October 30, 1991

Better get under cover, Sylvester. There's a storm
blowin' up, a whoppah, to speak in the vernacular
of the peasantry. . . . Poor kid, I hope she gets
home alright.

—*Professor Marvel,*
The Wizard of Oz

Nearorth Beach is a world apart. Part ocean, part desert, part
marsh—it stretches south from the parking lot in Orleans to
the new inlet in Chatham. Indigenous species of toads, rabbits,
grasshoppers and skunks inhabit tiny thickets of vegetation that
thrive in the swales of the ever-shifting dunes.

By day, hollows in the dunes act like natural solar collectors.
The sun reflects off their parabolic sides, concentrating its rays on
the floor of the depression. In the summer, dune sands reach foot-
scorching temperatures that rival those in many deserts and mold a
singular assortment of plants and animals.

It is the beach grass that holds this fragile world together.
Their unseen, underground latticework of roots anchors the spine

of shifting dunes that protects the bay behind. Often the system fails. Winter storms and moving sands have carved out a variegated landscape of blowouts, hollows, swales and washovers. These in turn have been colonized by beds of wild roses, thickets of bayberries, pockets of reeds and the seemingly fragile vines of morning glories.

It is also the habitat for numerous endangered species. Piping plover and least terns nest in new washovers and roseate terns on new sand islands. Equally endangered are the humans who call this stretch of sand and stunted vegetation home. They live in the small clusters of camps that snuggle into the flanks and hollows of the dunes. The Edsons lived in what was unofficially called "Backlash Village," named after one of the oldest camps. The name comes from the nasty habit of old-fashioned surf-casting reels to run free and create an unimaginable tangle of fishing line that can ruin a good night's fishing.

By custom, when North Beachers are in their camps they fly a battered American flag to signal their presence. On October 30 only two flags were flying, those of the Edsons and of Patrick O'Connelly.

Donna Edson hurried through the sandy twin tire tracks that cut through a low swale behind the dunes. She loved this spot. During the summer the swale was filled with a copse of phragmites reeds that rustled in the quiet breezes just above her head. The patch of phragmites was the closest thing to a forest that North Beach could offer. It helped quench her thirst for green grass and trees. She had remembered reading somewhere that long-distance sailors craved the sight of green vegetation almost as much as the sight of women, and she understood the craving.

She paused to listen to the long, reedy stalks as they jostled in the raging winds. It was a scratchy, ethereal, frightening sound and it matched her restless mood.

She was feeling nervous and edgy. It was time to move on, to leave North Beach. She felt the same nervousness that she imagined the terns felt before they started their long migration southward.

The week before they left for good, Donna would watch the entire flock rise up and settle down again at the slightest provocation.

It was time for the Edsons to migrate as well. Donna looked forward to the green grass of Florida and the four orange trees that she and Bruce talked about as if they were their children.

Living on North Beach was an acquired habit but it was one that Cape Codders had been cultivating for generations. The marriage of the Edsons had united two old Cape Cod families, the Mayos and the Nickersons. Donna had grown up in her father's old camp, "Fort Mayo," and Bruce had inherited forty acres from his Nickerson grandmother.

The old families had bought the land from the Indians so that they could graze their cattle in the extensive meadows of salt hay behind the dunes. Later the families build small camps for shelter during the fall hunting season.

In recent years the camps had become popular during the summer. They were places to sleep after spending all night surf casting and were blessed retreats from the bustle of the mainland. Most of the North Beachers retained an ingrained distaste for crowds and regulations. The irony was not lost on them that, with the coming of the Cape Cod National Seashore, they were now living in one of the most regulated spots on the globe.

The Edsons had been migrating from their camp on North Beach to their mobile home in Florida ever since Bruce had lost his job as a senior pilot with Air New England. The local airline had crashed financially, resulting in a change of its colloquial name from "Scare New England" to "kamakazi airlines." Always the captain, Bruce still rose every morning at 5:30 sharp to catch the tide or avoid the noonday sun.

Bruce had become the unofficial mayor of North Beach. He kept everyone's camps in repair, made a living shellfishing along the backside and watched over the flats in his role as weekend deputy shellfish warden. Many had suggested that it was like putting the fox in charge of the chickens, but Bruce had more or less risen to the occupation.

Bruce loved the wild beauty of North Beach and only visited the mainland with reluctance. In contrast, Donna liked to get off the dunes. She drove down the beach every day in order to work in a lawyer's office in Chatham. "It feels good," she said, "to put on high heels every now and then."

By October 30th the wind had been blowing steadily for three days. The Edsons weren't concerned. They had lived through many northeasters and the marine radio had reported nothing exceptional about this storm. Donna climbed to the top of one of the dunes to join Patrick O'Connelly.

Below them, twenty-foot waves assaulted the face of the dune. They had to yell to be heard above the pounding breakers. As the tide rose, the waves came higher. Finally one wave split the Holleck's camp in two. Half of the camp stayed in place, while the other half was carried into the middle of the beach. Another wave undermined the Lund camp and it was carried out to sea. Donna couldn't watch the third camp break apart and be washed away.

"I have to go back and help Bruce move the cars into the dunes. Why don't you have supper with us tonight? It's too damn scary to be out here all alone."

Dinner was short. It was just too exhausting to yell over the roar of the surf. The hurricane lamps flickered and the camp shuddered, but the three diners were still not concerned. It was just another northeaster that would blow itself out in three or four days.

Finally they abandoned the pretense of talking and eating. Patrick thanked them and promised to light his lantern so they could see that he had made it back to his own camp, which was only a hundred feet away. They watched the back of his yellow rain slicker as he trudged through the howling wind and lonely dunes. Within minutes the slicker disappeared in the gray wall of rain and spray torn from the crests of the raging surf. At last Donna saw the tiny flicker of Patrick's lamp. He had made it back. Exhausted, the Edsons fell into bed.

October 31st broke warm and gray. The wind was still up and an ominous tropical warmth filled the stormy skies. During the night Hurricane Hazel—packing terrible winds—had unexpectedly backed into the northeaster. At the same time, the remains of Hurricane Grace blocked the combined forces of Hazel and the northeaster from moving offshore.

Bruce arose at his usual hour and went outside. Donna was still exhausted from three days of shouting over the wind. She wanted another hour of sleep before driving down the beach to work.

"You gotta get up and see this."

Donna pulled on her robe and rushed to the door. "Oh, my God!" she exclaimed.

All the Edsons could see was water. Most of the dunes had been flattened, most of the camps had disappeared. Water was everywhere. It wasn't just water from a cresting river or water from a stormy season. It was the Atlantic Ocean, and there was no guarantee that it would subside.

The front of their Cherokee was in the water and the back of the GMC truck was buried in the sand. Donna rushed inside to call her daughter on the cellular phone they used every morning to keep in touch with the mainland.

"We're fine and Patrick's flag is flying, but you wouldn't believe it. The camps are all gone, all the vehicles are tipped over."

"The Coast Guard just called to ask if you're all right. They said they're too busy salvaging their boats and dealing with the harbor to get to the beach unless it's an emergency. Do you want to get off?"

"We'll be okay. Bruce'll get the truck started and launch the boat. He ought to be able to make it over to Ryder's Cove."

It took an hour to start up the vehicles and get them out of the sand. Bruce hitched them both together and managed to haul the twenty-foot workboat out of the dunes where they had left it for the night. After a few tries, the outboard started and Bruce headed across the bay. Soon eight-foot seas were slopping over the stern. He threw the anchor overboard and swam back to shore.

Dripping wet, Bruce and Donna huddled in the kitchen, watching the boat as it bucked in the waves and dragged its anchor. "It's goin' to sink right in front of our eyes," muttered Bruce.

Meanwhile, a rescue effort was being launched on the mainland. Skip Stearns had contacted a team of North Beach camp owners. He would lead a caravan of beach buggies down the beach at low tide. When the caravan reached the beach, however, Deputy Chief of Police Wayne Love told them he would have to turn back, explaining he couldn't risk losing the town truck. "Now, I'm not goin' to stop you guys, but fer Christ sakes be careful." The caravan pressed on.

It was like seeing the cavalry when Donna finally spotted the line of beach buggies threading it's way down the ravaged shore. The buggies had to stop in the high dunes and wait while Bruce and Donna waded through waist-deep water to reach them. All Donna was able to carry out was her pocketbook and their toy poodle "Jingles."

Bruce was nervous the whole way back. Always the captain, he was used to driving himself down the beach. The three buggies had to race the rising tide. They drove fitfully, stopping and starting between the base of the dunes and the Atlantic Ocean. They would race down the beach while one wave receded, then drive up the face of the dune to wait while the next wave lapped at their wheels.

By late afternoon, the three beach buggies, with their exhausted occupants, had made it back to the mainland. Of the 17 camps in "Backlash Village," only 3 had survived—the Edson's, Patrick O'Connelly's and one more. Florida would look especially good this year. The Edson's loved being so near the Gulf of Mexico.

"You see, we just like living on the water," explained Donna.

CHAPTER 18

The Halloween Storm: Chatham Fish Pier

October 30, 1991

Everything's just fine; look at the houses floatin' by.

—Wharfinger Stuart Smith,
comment to a journalist,
October 30, 1991

The day of the storm was different at the fish pier. By lunchtime Stuart Smith knew it was going to be bad. The manager of the wharf gazed out the window.

"Look at it out there. Should be dead low but the water's almost lickin' the top of the pier."

The wharfinger could not remember a time when there had been no low tide. It was supposed to be low at one o'clock, but the water was still at mid-tide levels as storm-tossed waves battered his pier with growing intensity.

"Wind's been blowin' steadily for three days and I got two big boats and a smaller one tied to the pier. Worst place to be in a storm like this."

All morning long the marine forecasts had grown steadily worse. The northeaster that was supposed to move offshore had stalled, blocked by the remains of Hurricane Grace further to the East. Now the combined storms were backtracking ominously toward Chatham.

Worried fishermen and their wives started filling Smith's office by mid-afternoon. Their hot coffee and help with answering the phones and moving boats was heartily welcomed. The memory of Hurricane Bob was on everybody's mind. As they watched, the ocean started to spill out onto the parking lot, the lights flickered and failed.

"Damn, juice's gone again. Look at that water; it's starting to come over the pier."

The wharfinger's office cleared out as fishermen scrambled to move their pickups to the upper parking lot. Suddenly Smith had an ugly thought.

"Jesus Christ, if the power's gone I don't know if the fuel lines will shut off. We could have diesel all over the harbor."

A tuna fisherman offered to help. Smith couldn't remember his name but he was sure glad to have his burly company. They stripped off their clothes and swam out the front door. The entire building was surrounded by water. They had to swim around the building to the front of the wharf where the boats usually tied up. They managed to climb into the unloading area that was knee deep in swirling green water. The tunaman held onto the edge of the platform while he caught his breath. "Let's wait here for a break in the waves, then dive down to take a look."

"Never in a million years did I think the water would get up to the fuel lines," Smith yelled over the roar of pounding waves.

"Here I go!"

The tunaman plunged into the frothing maelstrom and dove straight underwater. He felt his way down the pier until he found the wooden cover and managed to wrench it off by brute force. The cover bobbed to the surface, followed by the beaming fisherman.

"Got that sucker!"

Together the two dove again and managed to shut off all the

fuel lines. Slowly they swam back to Smith's office, avoiding the skiffs and dinghies careening wildly about the flooded parking lot.

Back in the wharfinger's office a routine had been established. People would huddle inside, drying off and drinking coffee, until someone spotted another boat that was breaking loose. Then everyone would race outside to try to secure the vessel as best they could.

The *Aaron Sarah* broke loose, floated over the pier and into the parking lot. All the pilings were underwater, so two fishermen had to tie the boat to a telephone pole while her captain put her in gear and maneuvered her back out of the parking lot. Su-Ann Armstrong was worried. Chris had climbed into the boat's rigging and no one could see him for half an hour.

"The Coasties are doin' a good job tonight," shouted a fisherman in a billowing sou'wester.

"Yep, they're earning their pay all right."

In the past, there had been some bad blood between the Coast Guard and fishermen, but all that was forgotten in the storm. There were 15 Coast Guardsmen saving boats, directing traffic and generally being damn helpful if not downright courageous. Every hour or so the "Coasties" would pack all the captains into the Coast Guard's 28-foot inflatable and take them out to their vessels to check on hawsers and bowbits. Earlier in the evening the Coast Guard had put their 44-footer out on the big offshore mooring. Three young "Coasties" spent the night on the bucking boat to ensure that she would be ready for any offshore mishap.

By 6:00 the wharf was shaking with the impact of every wave. Smith shuddered with every hit.

"I hope they don't knock the building her off her foundations."

"How's things going?" asked a young reporter from one of the local newspapers, who had just arrived.

"Just fine; look at the houses floating by."

Each wave seemed to be carrying torn shingles, doors or whole sections of walls and roofs.

Around midnight Nick Brown's *Synergistic* floated into the North jog and fetched up against the Coast Guard's inflatable boat. An exhausted Coastie helped Nick re-secure his lines. "There, now you've got a $140,000 bumper to fend you off."

At about 2:30 A.M., Bobby Ryder couldn't sleep and came down to the pier to check on his boat. He and Nick Brown walked along the North jog, with Bobby looking into the water. "Guess the camps aren't doing too well."

"Why not?" asked Nick.

"That's my kitchen table over there on the rocks."

Ryder's camp had been one of the northernmost of the camps that nestled in the dunes of North Beach. It had been blown down and swept almost a mile across the harbor. He hated to think what had happened to the thirty or so other camps on the outer beach. Little did the two know that, as they spoke, Donna and Bruce Edson were huddled together in their camp, wondering if they would survive the night of howling winds and crashing surf.

The next morning the wind was still up, but Stuart Smith was able to start to tally the damages. Five of the big boats were completely destroyed. Many were worth up to a quarter of a million dollars. Debris from the North Beach cottages littered the shore and propane tanks that had been ripped from the cabins lay in the harbor like unexploded mines. By late afternoon, Stuart learned that the Edsons had been rescued at North Beach. By that evening the winds had abated.

A few days after the storm, Cecilie Brown answered the phone. "It's for you, Nick. It's Su-Ann Armstrong."

Nick had always liked Su-Ann. She was tall, attractive and blond. He figured she could have easily been a model. But she had proven herself by fishing alongside her husband for several years before the birth of her three children. Su-Ann came right to the point.

"Nick, you have an economics degree. Won't you help us pull some numbers together to fight the Corps?"

It was hard to turn down Su-Ann Armstrong. "Okay, Sue. Drop the papers by and I'll take a look at 'em."

Nick Brown spent all that night poring over the 300-page feasibility study. He was astounded at the inaccuracies in the document. The next day he started preparing his rebuttal.

The day after the storm, Bart Nelson called his father. "Dad, that artificial sand dune the state made us build didn't do so well. It washed away in fifteen minutes."

"That was $30,000 worth of sand. Now maybe they'll listen to us. How about the house?"

"Not so great, Dad. Stu Crosby and I had to shovel four feet of sand and water out of it. A lot of the electrical system shorted out and there's only about 19 feet of beach left in front of the house. I think we should call Nick Soutter again."

PART III

Tying up Loose Ends 1992–1994

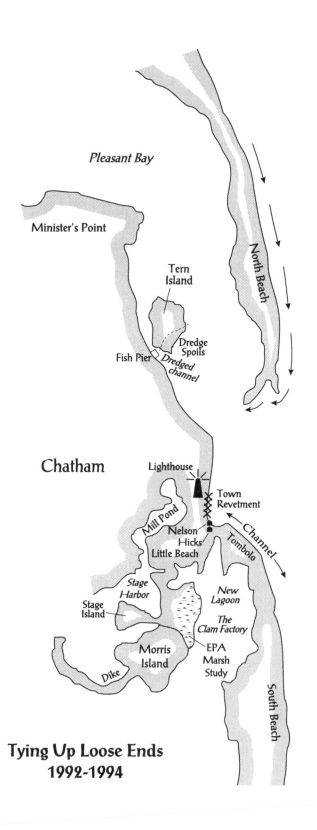

Pleasant Bay

Minister's Point

Tern
Island

Dredge
Spoils

Fish Pier *Dredged channel*

North Beach

Chatham

Lighthouse

Town
Revetment

Mill Pond

Nelson
Hicks
Little Beach

Channel

Tombolo

Atlantic Ocean

*Stage
Harbor*

*New
Lagoon*

Stage
Island

*The
Clam Factory*

EPA
Marsh
Study

Morris
Island

Dike

South Beach

**Tying Up Loose Ends
1992-1994**

CHAPTER 19

Orrin Pilkey

February 29, 1992

It's awfully tempting at times to take what I've
learned in North Carolina and go galloping off in
all directions.

—*Orrin Pilkey*,
Oceans *magazine*,
April 1987

The Halloween storm had left its mark on Chatham. It had
caused four times as much damage as Hurricane Bob. The
Fish Pier was left standing, though some of the tie rods that held
the bulkhead together had popped loose. Most fishermen thought
the building would have been knocked off its foundation if the
storm had continued through another tidal cycle.

A tenth of the fishing fleet was out of commission for the first,
long, lean months of winter. But the biggest blow to fall on the
fleet was the official release of the Army Corps of Engineers' report. As had been feared, Chatham was turned down flat, threatening the jobs of 500 people and the loss of thirty million dollars
of yearly community income.

The storm had destroyed fourteen camps on North Beach and
damaged forty houses on the mainland. Eighty more homes on

Little Beach were now in jeopardy. In all, Chatham had sustained $750,000 worth of damage and faced $60 million dollars in future losses to it's homes and fleet.

However, the storm had also helped break up a log jam of indecision. The town had become familiar with the plight of people who were losing their homes. It started to get easier to receive permission to build seawalls. People took notice when the Conservation Commission gave the town quick approval to build a $750,000 revetment to protect the Chatham Lighthouse. After opposing revetments for many homeowners, the town was now building the biggest seawall of all.

Chatham had also put together a coastal erosion advisory committee in response to dissatisfaction with many of the town's past decisions. The committee was about to hire Dave Aubrey to put together an Environmental Impact Review for the town's entire eastern shore. It was the same EIR that the state had requested that homeowners put together in 1987. As part of the process, Dr. Aubrey would also prepare a 50-year plan to deal with erosion that would be caused by the new inlet. In short, the town was finally pulling itself together to come up with a common-sense, flexible plan to deal with its erosion problems.

It was against this background that The Friends of Chatham Waterways invited Orrin Pilkey to speak on a blustery, frigid day in late February. Two hundred and fifty people braved the blowing snow and jammed into the Chatham High School gymnasium.

Orrin Pilkey was an affable, witty speaker who enjoyed stirring up the crowd and making provocative statements. He stood 5 feet 4 inches tall, weighed 220 pounds, dressed sloppily, and sported an unruly salt-and-pepper beard. He sounded like a prophet with a Ph.D. Some said he preached the gospel of coastal destruction.

Pilkey had become famous for leading small armies of journalists down beaches that had been recently devastated by coastal storms. Newsmen knew that he was always good for a quick quote whenever a hurricane was cruising up the coast. He was so happy to oblige that producers for ABC's "Nightline" had dubbed him "Mr. Soundbyte."

The media loved his message. It was short, simple and carried the authority of science, but with a tinge of biblical certainty.

Pilkey advocated that homeowners who were faced with sea level rise should simply move. He was the most outspoken and influential proponent of this school of thought. His students from Duke University had landed jobs in the bureaucracies of many coastal states and had been influential in rewriting their coastal policies. However, some of his students admitted that it was easier to make coastal policy in Pilkey's classroom than in the real world, because "he doesn't have to deal with people crying the day after they have lost their home."

In his writings, Pilkey had advocated bulldozing all structures within 300 feet of the coast. That would have effectively obliterated the Chatham Fish Pier, the Chatham Lighthouse, over a thousand homes, hotels, restaurants, and half of Chatham's tax base. He did not reiterate the idea in Chatham.

Dr. Pilkey opened his talk by showing a slide of St. Catherine's Island off the Georgia coast. The beach was obviously scarped, but Dr. Pilkey stated that there was no erosion problem there because there were no buildings. One homeowner whispered into his neighbor's ear, "That's like asking if there's any noise if a tree falls and nobody hears it."

Dr. Pilkey described three options for fighting erosion when it threatens buildings: seawalls, beach renourishment and coastal retreat. He stated that he would like to see all ocean seawalls made illegal. He felt that seawalls destroy beaches and that "sooner or later the money spent on building seawalls will be wasted when the walls are ruined by the implacable force of the ocean."

"But none of the 24 people who have been given permission to build seawalls have lost their homes and the eight people who were denied seawalls have had to watch their homes be washed away," commented a homeowner.

"Well, I hope they were decent sports about it. You see, you and your kind are a small number out of all the people who would like to use the beach." Some in the audience applauded.

"But Chatham has just gained South Beach. That's a mile and a half of one of the best beaches on the East Coast," continued the homeowner.

Another member of the audience asked, "How long is 'sooner or later'? We might only have to wait ten to twenty years for North Beach to grow long enough to protect us again."

"Well, most seawalls break down in a matter of decades."

Dr. Pilkey recommended that people consider beach replenishment, but admitted that it was expensive. Albert Nelson stood up to agree. "The state forced me to renourish my beach, and I watched $30,000 worth of sand wash away in 15 minutes."

Dr. Pilkey's favorite solution was coastal retreat. Some of the audience loved the message. They saw him as a prophet, bravely cleansing the beaches of America by sweeping away coastal homes. In one of his papers he had described sea level rise as a "golden opportunity" to turn private land into public beaches.

Others were skeptical. "He makes sea level rise sound like the inevitable march of progress. It reminds me of the arguments of the early urban renewal enthusiasts who tore down whole neighborhoods and displaced thousands of people in the name of progress."

After the meeting, part of the audience drove to the Squire to decompress.

CHAPTER 20

Back at the Squire

February 29, 1992

> It's not fair that Pilkey can make a speech, pick
> up his honorarium and head for the next town
> destroyed by a hurricane. It's not fair that I have
> to stay behind paying lawyers outrageous fees in
> order to get some money for the house I lost
> because of Pilkey's theories.
>
> *—A Homeowner,*
> *February 29, 1992*

A fter Pilkey's talk, several homeowners, their lawyers and consultants retired to the Squire. "He sure can charm an audience," observed one lawyer to another.

"But you know what bothers me? We witnessed a very subtle form of social cruelty. Pilkey knew that most of the audience was well-educated, articulate and didn't have anything to lose. But when some poor guy who is about to lose his house speaks up, Pilkey says, 'you'll just have to be a good sport about it.' The guy's wife gets upset and mispronounces Pilkey's name. Pilkey makes a joke about it and everybody laughs at the couple's expense." That's just plain mean in my book."

"But I think he makes some of his provocative statements just to make people think. In private, I've heard him be very reasonable and admit that sometimes a seawall is a good idea. But once the cameras start rolling, he'll deny it to his last dying breath," reflected the lawyer.

"Perhaps he sees himself as a foil to the engineers. They're always saying there's a solution, so he feels he has to say there isn't one." The biologist got up to buy another beer.

"Of course, it's always easier to be on the side of the angels than to solve real problems. In the short term the engineers are right, but in the long term Pilkey's right."

A geologist scratched his beard and observed, "You know the thing that interests me about engineers is how they worship their ancestors. They'll never criticize some famous old engineer who made a mistake. In science, young turks are always trying to knock one of their predecessors off his pedestal. It's not very pretty, but it's probably more intellectually honest."

"Of course, Pilkey's position is becoming mainstream. Just because we've had so many storms, everyone is thinking that sea level rise has increased. But even if the sea level had been falling, we would have had this much erosion," observed his companion.

"Of course that's it. Storms are the real problem, but the press laps it up if you say that the coasts are going to disappear because of sea level rise." The geologist stared at the writer in the crowd.

"Well, you know journalists have their own traditions, too. We've been trained to find people with opposing sides of an argument. In science, that can be very misleading. Everybody in the field might agree that something is so, but the press will give equal weight to some crackpot with a totally unbelievable idea. That's okay in politics, where everything is a matter of opinion, but in science we know that some things are true and other things are not."

The geologist sipped his beer and observed, "Well, I think it was a pretty interesting talk. But you know it's nothing new. I remember back in the '50s there was an old guy who lived on the Outer Banks in the Carolinas. It was rumored he was a drunk and he'd come to all these meetings and say the beach was migrating and everyone'd laugh at him. Of course, it wasn't til the '70s that scientists got around to discovering that he was right after all."

A homeowner piped up, "You geologists all think alike. You get all excited when you see some change because you see it so seldom. Who cares if we have 6 inches of sea level rise in the next hundred years? We don't live in geological time, we live in human time. We don't care about a hundred years, we care about whether we can live in our house for the next ten years or whether we can sell it in twenty years."

"Yeah, but 'let's start thinking about tomorrow.'" The geologist hummed the tune with a twinkle in his eye.

"Give me a break. Do we think that far into the future in any other sphere of human activity? Will you vote for Clinton because Dan Quayle's grandchild might get elected next century? Will you sell your family store because we might have a depression in the next century? Will you move to another country because we might be invaded in a hundred years?"

"Sure, I'd vote for Clinton to avoid Quayle's progeny!"

"Well, maybe that wasn't such a good example. By the way, have you read Al Gore's book?"

"Yeah, thought it was pretty good."

"Did you like that part about when he was a boy learning to stop erosion before it gets started real good?"

"You're not goin' to catch me there. He wasn't talking about ocean erosion."

"No, but a lot of homes in Chatham could have been saved and a lot of people could have spent a lot less money if they had been allowed to build seawalls right away."

"What did you think about his global Marshall Plan to save the environment?"

"Not worth a pinch of coon shit. Politicians love to look at things from the top down. We biologists have great faith in messy, inefficient systems that work from the bottom up. Who would have predicted that tiny strands of DNA would create an iris or a whale? It all came from small yes-no decisions throughout evolution.

"The same thing goes for human systems. The reason that democracy or the market system work so well is that they rely on lots of little yes-no decisions that filter their way up to the top. I don't think you are going to save the world from the top down, but by making lots of little yes-no decisions from the bottom up. We don't have enough information yet to know if our top-down policies are the right ones."

The biologist's colleague joined in, "We do know that a lot of top-down solutions to global warming will be incredibly expensive. To cut back on CO_2 enough to slow down global warming would mean that the entire world would have to use the amount of energy it used prior to World War II. Can you imagine the disruption that would cause? You would have to give up your television, your VCR, your computer, half your car and most of your job."

The geologist cut into the biologist's long monologue. "It might just be a case where the cure is more expensive than the problem. I certainly think we can deal with the effects of sea level rise, but I'm not sure we can prevent global warming if it's really going to happen. It may be one of those cases where we just have to muddle through as best we can."

"Ah," sighed the biologist, "that might be a little hard for humans to accept. We're an arrogant species, you know. I think we secretly take pride in the fact that we are powerful enough to screw up the planet and are the only ones that can save it. We don't like to think that we can't control the future."

"Of course, we don't even know if CO_2 is really the problem. All the measurements of increasing CO_2 come from instruments at the top of Mauna Loa in Hawaii. On the marsh, you might find four times as much CO_2 a centimeter above the peat, twice as much a meter above it and half as much in the middle of the day when photosynthesis is really cooking."

Two graduate students wanted to join in on the fun. "Methane is more potent than CO_2. If you really want to slow down global warming, stop eating meat. We're deforesting rain forests to raise cattle for MacDonald's hamburgers.

"Or start to eat pork. It takes less land to raise pigs. Besides, they fart less than cows. Less farts, less methane."

The woman behind the bar offered a drink to a homeowner and observed, "You homeowners are just a bunch of masochistic Puritans; you enjoy having to rebuild your houses every year."

"Puritans! Hell, we're the Indians. Our lifestyle has just become old fashioned. The town thinks it can make more money on tourists.

"I loved my home on the beach. I loved to get up in the morning and see the sun rise across the ocean. I loved to take my kids fishing and to scratch for quahogs at low tide. I loved to share a beer on the beach with friends while watching the sunset.

"I probably looked out the window of my house at least a hundred times a day and every time I saw the ocean it gave me a jolt of pleasure. I built my life around living on the ocean and they have taken that away from me."

The old geologist interjected, "But you can't fight sea level rise."

"Of course you can. Look at the people who were allowed to build seawalls. They're okay, aren't they? Besides, I'm not talking about sea level rise; I'm talking about making it through the next ten or twenty years.

"You're not talking about sea level rise, either. You're talking about banishing me from my home, ripping me away from the thing that I love the most. Something that has . . ."

"Can I say something?"

"No, shut up, science head. I'm not done yet. You're not talking about sea level rise; you're talking about money and politics. You don't want us on the beach so that tourists can boost the economy. You want 'em to pay for a motel, spend a few hours on the beach, then go to the shopping mall or multi-plex cinema. That's the Cape Cod you want."

"That's not fair."

"Sure it's not fair. It's also not fair that Pilkey can make a speech, pick up his honorarium and head for the next town destroyed by a hurricane. It's not fair that I have to stay behind and pay Nick Soutter outrageous fees to try to get some money for my home. It's not fair I lost my house because one of Pilkey's students drew up some cockamanie regulation out of a textbook. I lost my home because of his regulation, not because of sea level rise."

A strained silence fell over the bar. A companion quieted the former homeowner as one by one the patrons drifted into the snowy evening.

CHAPTER 21

Legalities

Spring 1992

A free press plays an essential part in the delicate
balance that exists in this country between
individual liberties and the power of government.
For unchecked government power . . . can com-
promise, even crush individual liberties.

—*Jack Nelson,*
Los Angeles Times, *1984*

April 30—Spring was still evolving as I sat inside our house
casually channel surfing between CNN and the weather
channel. Like many Americans I became hooked on CNN during
the Gulf War. Now it looked like our troops would be diverted to
Bangladesh.

A tropical cyclone had swerved out of the Indian Ocean and
had swept up the Bay of Bengal. It packed a four-foot storm surge
and twenty-foot waves—the same size waves that had pounded
Chatham during the recent Halloween gale. However, in
Bangladesh over three million people live on land less than a meter
above sea level.

The results were staggering, the numbers almost beyond comprehension. Thousands of people had been swept out to sea, never to be seen again. Many had been devoured by the marauding bull sharks, but CNN happily spared us from those gruesome details. Thousands more people were left stranded and died from lack of food and water and from disease. In the end, over 140,000 people had died and 10 million were left homeless. There were fears that millions of environmental refugees would soon be migrating into neighboring countries, although just the reverse occurred. After the cyclone passed, the major concern of most Bangladeshis was to return home before their houses were claimed by their neighbors. The CNN anchor people kept repeating that they just couldn't understand why the Bangladeshis would return to places subject to such devastation.

I put down my clicker. I knew it might seem ludicrous, even obscene, to compare the problems of Bangladesh with the problems of people on Cape Cod. However, in my bones, I knew I had learned something important and elemental during the last few months. I knew I would fight to the finish to protect my home and I knew that if I lost that fight I would suffer greatly. It stemmed from a deep, strong, biologically based territoriality that is so often forgotten in debates about sea level rise. There is a universal and fundamental characteristic among humans that they will fight to the finish to save their homes, whether in Bangladesh or on Cape Cod.

In even the best of times, the lives of Bangladeshis are hard, cruel and poverty stricken. However, Bangladeshis know something that we tend to forget. Their land is immensely fertile, their coastal waters are some of the richest in the world, and the shores of the Bay of Bengal are their home. To lose the place where they live would mean poverty, banishment, homelessness—a human tragedy that would reverberate through their families for generations.

How could the government of Bangladesh prevent people from returning to their homes? It was clear they could not. Tropical cyclones had ravaged this coast for hundreds of years, and yet after every storm people returned. To the Bangladeshis the benefits of living along the shores of the Bay of Bengal far outweighed the risks. Their government is simply not strong enough to overcome this most basic of human drives, the desire to reestablish one's home.

But what of our government? How does it deal with controversies that pit environmental regulations against human liberties? Let's return to my family's coastal bank.

We had won permission to build our gabions. The stakes that had skewered me during our first go-round with the state helped us in the second. Several more storms also helped. We were given permission to reapply.

This time the stakes helped. They showed that since the inlet had opened the rate of erosion was five feet per year instead of six inches. Under new guidelines suggested by the National Academy of Sciences, five feet per year put our house in imminent danger. It had cost us $10,000 to pay for fees, engineers, lawyers, and environmental consultants, but we had finally obtained permission to build a $20,000 gabion.

Others had not been as fortunate. A dozen people had lost their homes on Harding's Lane in Chatham, twenty more people had lost summer camps on North Beach, and 120 houses were in jeopardy on Little Beach, and on Morris and Stage islands.

Others had gone through the regulatory mill. Bert Nelson owned the first house on Little Beach. It was an unfortunate location. His house stood where the coastal bank becomes a dune. The Wetlands Protection Act states that you can build a seawall in front of a coastal bank but not in front of a coastal dune. This meant that his neighbor, the Chatham Beach and Tennis Club, only six inches away, could build a seawall to protect its tennis court but Bert couldn't build anything to save his home.

I walked along the beach in front of the two neighbors to see if I could tell the difference between the bank and the dune. I could not. In fact, most geologists couldn't tell the difference, either. The way the state distinguishes between a bank and a dune is to put some sand grains under a microscope and compare their size.

If they are small, it's a bank; if they are large, it's a dune—which seems like a pretty fine distinction if your house is on the line.

But Bert had another problem. His neighbors' seawalls were starting to erode the beach that protects his house. Bert couldn't build anything to protect his home but he had been given permission to renourish the beach with $23,000 worth of sand. It had washed away in fifteen minutes during the Halloween gale.

After the storm, he had been given permission to do one of several things: He could renourish the beach again with another $100,000 worth of sand. He could spend another $100,000 to move his house back thirty feet, to a location where it would be considered to be on a coastal bank, so then he could obtain permission to build a seawall. Or, he could build a temporary sandbag seawall for $75,000. The theory behind the temporary wall was that it would buy him time until an appropriate Environmental Impact Review could be completed. Of course Bert would have to pay $30,000 to get the review started, but that would eventually be paid back by the other homeowners on Little Beach. It seemed like a strange way to manage a coast.

I started to think the unthinkable. What if the numbers are wrong? What if we have put together regulations that are based on estimates of sea level rise that are incorrect?

The Wetlands Protection Act was cobbled together in 1978. At the time, climatologists were predicting that global warming would cause a catastrophic rise in sea level. Dr. Stephen Schneider had published a paper in 1980 that showed that global warming would cause the sea level to rise by 28 feet in the next 100 years. The media had scooped up the report and had reprinted it with flashy graphics showing New York and Washington, D.C. under 30 feet of water. Much less attention was paid when Dr. Schneider quietly retracted the numbers in an obscure footnote in *Scientific American* in 1989.

Since then, the estimates of sea level rise have steadily dropped from 28 feet to 12, to 9, to 6, to 3, to 1 foot during the next century; approximately the same as during the last century.

So what happens when public policy is based on numbers that are wrong? Individuals may get hurt. We saw this during the '60s and '70s when old neighborhoods were torn down to make way for urban renewal. In the '80s we discovered that those stable old low-crime neighborhoods were exactly the areas that should have been saved. But by then it was too late.

Coastal Zone Management offices have been given some of the same broad police powers as those given to the Defense Department when it wants to build a new military base, or cities when they want to change zoning or bulldoze a neighborhood for urban renewal. However, in those cases the government is required to compensate homeowners under the 5th Amendment to the Constitution. Not so with coastal regulations. The state can say that you are not allowed to do anything to protect your house, and so you lose it, along with all your life's savings.

The question of the constitutionality of such regulations was clarified on June 29th, 1992 when the Supreme Court ruled in the case of Lucas versus the South Carolina Coastal Commission. David Lucas was a developer who claimed he lost the value of two shorefront lots when the South Carolina Commission passed a new regulation that denied him permission to build houses on his property. The Supreme Court upheld the Fifth Amendment to the Constitution that states that the government may not take a person's private property without just compensation.

When the decision first came out, many environmentalists thought it would throw a monkey wrench into managing the coasts and might well break the government treasury. However, if we are going to have only one foot of sea level rise in the next hundred years, coastal erosion is not going to be a major problem. After all, Chatham experienced the equivalent of fifty years worth of sea level rise in one night.

Many coastal towns have a few low-lying areas that get flooded during severe storms. If we simply use a little common sense, we can decide which houses can be saved and which should be removed. If we simply buy out every home that is destroyed after a major coastal storm, it would still be less costly than buying another F-16 fighter plane every year. It seems like that is a small price to retain a basic constitutional right of a liberal democracy.

However, the larger question is how much (or whether) we should abandon the traditions of a liberal democracy in order to cope with environmental problems. In some countries, the question might seem like a luxury, but it is not.

I was reminded of another scene. During the elections in the former Soviet Union, CNN showed a candidate of one of the new liberal parties. Behind him, a shield declared the principles of the party in three words: "Law, Democracy and Ownership." The juxtaposition of ownership with law and democracy struck me as odd, almost jarring.

I suppose we take ownership so much for granted that we tend to think of it as a conservative, slightly embarrassing part of the soft underbelly of capitalism. It took the elections in a former communist country to remind me that the right to own property—whether it is your house, your land, your farm or a small business—is in fact one of the very cornerstones of a liberal democracy—a crucial right that must be preserved to counter the overwhelming power of government.

Perhaps our forefathers knew more than we sometimes realize. There is nothing quite like the prospect of losing your home to make you appreciate the wisdom our Constitution can provide in times of crisis. Maybe there's hope, when even an aging child of the Sixties can still learn a few lessons about our form of government.

CHAPTER 22

The Marsh

June 1, 1992

Spring is a time for the renewal of life. In a
sudden reawakening, incredible in its swiftness,
the simplest plants of the sea begin to multiply.

—*Rachel Carson,*
The Edge of the Sea

Spring had its first tenuous grasp on the land. The May sun had
thawed the marsh, but winter still lingered in the cold waters
of Pleasant Bay.

The marsh was showing the first faint signs of spring's reawakening. A subtle filigree of green lined the creeks and shore. Gradually the verdant hues of the marsh grass would emerge to overflow
the creeks and flood the meadows of the upper marsh.

Jane Vollers was setting up a tripod to photograph the section
of the marsh where I had driven in stakes in 1987.

"I take photos of this section every week. Every few months I
send 'em out to Bob Hone on the West Coast who transfers 'em to
videotape. After several years we'll have a nice little time-lapse se-
quence of how the marsh has changed since the inlet opened."

Hudson Slay stood by quietly. He was one of Orrin Pilkey's
graduate students and was spending the summer as an intern for

the Environmental Protection Agency in Boston. I pointed to the stakes. "I put these out in 1987, expecting to show that the marsh was dying back because of the higher water. It was only this year that we realized something else was happening. See where the cordgrass has grown beyond the stakes? It seems to be invading the salt marsh grass about six inches a year."

"Hmm," said Hudson Slay, looking at Jane.

"Yeah, and that's where I saw that damn dead eel."

Not far away a small group of students was trying to work a 6-foot hand corer down through the marsh's peaty interior. Gradually they retrieved the shaft and gently laid a 6-foot core of peat into a shallow trough. The core had curious striations of gray, yellow and brown. The yellow band was a 3-inch layer of sand that was tucked between bands of loam and peat. The layers told a complicated story of sea level change.

At the edge of the marsh the story was being repeated. A long tongue of sand sat on top of the marsh. During the past few winters, the sand had eroded from a nearby bank and snaked its way along the shore. During the Halloween storm, it washed over the beach and flowed into the marsh.

One of the students scrutinized the area on hands and knees, and observed that "tiny shoots of beach grass are already sprouting through the sand."

I paused to explain. "If we could come back in a hundred years and take another core of the marsh, we would see several inches of loamy beach grass remains, and a six-inch band of sand lying on top of several feet of marsh grass peat. It would tell us that the level of the bay suddenly rose, erosion increased and beach grass started to replace marsh grass. We could guess it happened because of the opening of the inlet.

"The marsh that we are standing on started growing about six thousand years ago. As the sea level rose, the marsh grew with it, building up thicker and thicker layers of peat. If we were to core down through the entire marsh, we would remove an 18-foot core of peat that represents the amount the sea level has risen in the past

5,000 years. In that record we would see abrupt changes every 140 years or so. They show when the inlet opened and abruptly increased the tidal ranges in Pleasant Bay, causing the marsh to grow more rapidly." The students seemed bored with my lecture. It was time for show and tell.

"Now this is *Spartina alterniflora*, or cordgrass. It's the first marsh grass that can colonize a shallow-water area. It is a fascinating plant that has evolved a unique mechanism that allows it to dominate the shores of estuaries." We bent down to look more closely at the fronds of the grass.

"See these tiny crystals on the blades of the *Spartina?* They are actually crystals of salt that the grass excrete through pores at low tide. This removes the salt that would normally kill the plant's cells."

"The *Spartina alterniflora* also sends out rhizomes, thick underwater root systems that catch mud and detritus to build up a foundation of peat. The build-up of peat allows the marsh to keep pace with the gradual rise of the sea level.

"As the marsh grows it also matures. As peat builds up it prevents the high tides from flooding the upper marsh. The *Spartina alterniflora* is replaced by its cousin *Spartina patens*, and a new environment is created, the dense meadows of what earlier settlers called salt marsh hay.

"Some biologists have stated that if the sea level rises more than half an inch for over three years, the system will break down. The marsh will not be able to build up peat fast enough, and the marsh grasses will die back because they cannot excrete salt at low tide. Obviously this would be disastrous to coastal areas and would seriously deplete major sources of protein in many areas of the world.

"Ever since the inlet opened, we have been looking for evidence of this die back but have not found any. What we have found is that high tides can once again flood the upper marsh and this allows the faster-growing *Spartina alterniflora* to again replace the *Spartina patens*. In essence, the marsh has changed from a more mature marsh to a younger marsh.

"However, the most significant aspect of this change is that the marsh and the bay have increased their productivity, because *Spartina alterniflora* produces about two-and-a-half times more biomass than *Spartina patens*. That increase at the base of the food pyramid reverberates through the food chain, increasing productivity at each trophic level.

"So, instead of seeing a calamitous die-off, we are seeing a powerful resurgence of life. The alterniflora grass is growing into the bay and into patens grass about six inches a year. So far the bay has gained about 34 new acres of cordgrass marsh.

"It is an unexpected finding that holds some important lessons for the future. It shows that natural systems have incredible powers of regeneration. There are many other examples of nature's ability to bounce back from change or destruction. Yellowstone National Park is recovering rapidly from the intense forest fires in 1988 and Washington is recovering from the eruption of Mount St. Helens. Often, in cases like these, destruction leads to greater productivity as more aggressive, rapidly growing species rush to recolonize a devastated area.

"Recently, scientists have been investigating species of plants that will grow faster as carbon dioxide increases in the atmosphere. These plants will not only increase productivity but will also consume more carbon dioxide to help reverse global warming. They will act as our biosphere's own immune system to help heal our wounded planet.

"Humans tend to be innately conservative. We assume that all the changes associated with global warming and sea level rise are going to be negative. However, the closer we look at individual ecosystems and the species that colonize and dominate them, the more we may discover that many habitats will not only adapt to the changes but that their overall productivity will rise.

"Humans might consider some of the species that trigger the increase in productivity to be 'weed' species, but their overall effect will be that the world, the biosphere, will respond quickly and positively to global warming. It is only rigid human agricultural practices that will suffer."

"Is it like the Gaia hypothesis?" asked a student.

"Yes, many of these changes were predicted by the Gaia hypothesis, the concept that the biosphere acts like a living organism to regulate itself. One of the early criticisms of the Gaia hypothesis was that it is too purposive. It seemed to require species to act altruistically and with some purpose in order to counteract changes in the environment. However, all the planet requires is a rich, diverse mix of species for the system to work. During times of change, some of the species will thrive and some of them will suffer. The ones that thrive will tend to counteract the changes and drive the system back toward homeostasis. It can all proceed according to the time-honored rules of Darwinian evolution."

The students seemed bored with my long, preachy lecture. But it seemed reassuring to ponder something that had such positive overtones at the end of a warm spring day when life flooded the bay with its abundance, and the marsh grew quickly and luxuriantly at our feet.

CHAPTER 23

New Lagoon

Summer 1992

The break was the best thing in the world for the shellfishery. It's certainly put a lot of people to work.

—*Stuart Moore,*
Shellfish Constable

A heavy, gray bank of fog rolled slowly across the Chatham clamflats. The deep groaning of the unseen lighthouse echoed eerily through the misty shroud. Occasionally a light onshore wind would tear a gaping hole in the mist, revealing the silhouettes of men and women hunched over shallow holes in the black anaerobic sand. Nimble fingers pulled glistening white, soft-shelled clams out of black sand and tossed them carefully into nearby baskets.

They did not pause to talk, but bantered back and forth happily without raising their heads. Only a few more hours of the high course tide were left and each clam meant a little more security for the coming winter.

A shellfisherman peered into Shareen Eldredge's wire basket. "How'd the skirt boat do today?"

"None of your goddamned business, peckerwood. Have a beer and get outa my space."

There was a happy gold-rush mentality on the flats. Many of the shellfishermen had been unemployed or had been making $5 an hour only a few months before. Others were putting themselves through college or raising capital for a new boat. The market for "steamers" was fluctuating between $60 and $90 a bushel, and some of the shellfishermen could make $250 for a half day of back-breaking labor. However, hanging over the flats was the realization that this bonanza could also end with another northeaster.

The fishermen were clamming in a new lagoon that had been created by the tombolo south of Bert Nelson's house. The tombolo had started growing as sand built up south of the new inlet. In essence, the northern portion of South Beach had migrated half a mile across the harbor and was now welded to the mainland.

The tombolo had created a productive new ecosystem that was protected from waves and was bathed daily with fertile waters from the Atlantic Ocean and Nantucket Sound. The conditions in the new lagoon were ripe for the resurrection of what the shellfishermen called "The Clam Factory," which had appeared so providentially in the '30s during similar economic and ecological conditions.

The new lagoon was supporting 400 commercial clammers and 40 full-time mussel fishermen, significant numbers in a small seaside community with a year-round population of 7,000 people and an unemployment rate of 8 percent. The creation of the new lagoon was like having Nissan call up the town fathers with an offer to build a new rear bumper assembly plant that would increase employment by 1 percent—nothing to sneeze at.

Chatham's clamflats had always been a natural safety net in tough economic times. To enter the fishery was easy and overhead was low; all you needed was a clam rake, a strong back and a commercial license available to any local resident for $200. Many fishermen bought one every year, "jist fer insurance's sake."

Hundreds of summer fishermen were also enjoying the large populations of lobsters and blue crabs that had increased since the opening of the inlet. But wasn't something wrong here? Hadn't scientists been warning Chatham officials that global warming and sea level rise were going to wreak havoc on coastal resources?

CHAPTER 24

Scientific Debates

August 1992

> Global warming means rapid change, and rapid
> change means the end of familiar places. It means
> that the creatures of Yellowstone will scatter and
> the languid salt marsh, with its clams and ducks
> and egrets, will be drowned.
>
> —*Michael Oppenheimer,*
> *"Dead Heat"*

In August, Hudson Slay released his report for the Environmental Protection Agency. It used Pleasant Bay as a case study for sea level rise and warned that decreases in salt marshes might "jeopardize the productivity of shellfisheries."

His study considered a small marsh that was eroding because it lay directly across the harbor from the inlet. The report concluded that sea level rise would decrease marsh areas because they would be "squeezed between the upland and the rising sea." An article of mine came out in *Earthwatch* magazine in 1993. It stated that

> If Dr. Slay had investigated Pleasant Bay as a whole, he
> would have discovered a strikingly different pattern. We
> have documented that over 95% of the marshes in the bay

are growing at the rate of 6 inches a year, and have added 34 new acres of fertile wetlands since the inlet opened.

The same story holds true for eelgrass, *Zostera marina*, the indicator species that is used to gauge water quality. Eelgrass beds in the new lagoon and lower bay have increased 1800%, adding 400 acres of new eelgrass beds that have helped attract mussels. In human terms this has been dramatic. In 1989 there were no landings of blue mussels in Chatham; by 1991, landings had jumped to 100,000 bushels and were supporting 40 year-round fishermen.

So what's the story here? Are global warming and sea level rise going to destroy our coastal resources, or does Pleasant Bay show that natural productivity will rise as marshes expand and new lagoons are formed?

Pleasant Bay does not entirely invalidate our concerns about sea level rise, but it should raise a healthy skepticism and renew our awe of nature's ability to adapt and thrive in an everchanging world. We should not be surprised—after all, that is what evolution is all about.

Particularly, we should not be surprised to see this kind of change in an estuarine environment. Every species in an estuary originally evolved to live in another environment, either on land, in the sea or in fresh water. Their success lies in their ability to rapidly exploit a new habitat and one of their strategies is profligate fecundity. Each adult clam can produce hundreds of thousands of eggs, so it is theoretically possible that all of the clams in the "clam factory" could have come from a single adult. It is probable that 90 percent of them came from less than a hundred adults.

It is because of the straightforward evolutionary history of these opportunistic keystone species that the bay has increased its productivity. However, one cannot help being intrigued by how providentially an invisible Gaian hand seemed to be working to return the bay toward health and stability.

Even so, there are some important distinctions to be made. Some of the changes in Pleasant Bay occurred because of a change in tidal range, not because of sea level rise per se. So a marsh that experiences sea level rise might still die back. This, in fact, was what I was looking for when I first started to study the marshes. This difference could account for the marshes, but not the dramatic increase in eelgrass and shellfish beds.

Another potential difference was the fortuitous creation of a new lagoon with excellent water circulation. This specific change might not occur in other areas undergoing sea level rise. However, other researchers have pointed out that lagoons are associated with sea level rise and that opportunistic colonizing species will take advantage of them. If sea levels were to start dropping, then we might see eutrophication and a rapid decline in highly productive lagoons and marshes.

Other areas of the world have already learned about the rapid increase in productivity after flooding. Two such areas are Bangladesh and Egypt, often cited as areas that are expected to suffer greatly from sea level rise. As CNN made so clear, three million Bangladeshis live on the dangerous deltas that encompass the Bay of Bengal and they are only one meter above sea level. Egypt is also expected to lose coastal land that accounts for 20 percent of its GNP.

However, increased productivity is one of the reasons that millions of Bangladeshi fishermen and farmers return to the deltas after typhoons have devastated the area.

It may be something in our human nature that leads us to assume that change will deteriorate the environment. However, ancient Egyptians overcame that predisposition and learned to pray for the flooding of the Nile, because they knew it would increase the fertility of their floodplain agriculture. Many of the modern calls for coastal retreat ignore the fact that many cultures have learned that coastal flooding can improve the natural productivity of coastal wetlands in a similar manner.

Many researchers also assume that sea level rise will affect all coasts in the same manner. In reality it will not. The entire coast will not recede the same amount. Low-lying areas will recede more than others, creating new lagoons that in turn will grow new marshes and increase productivity.

An additional way that sea level rise will manifest itself is through breaks in the system of barrier beaches that protect our coasts. After Hurricane Bob, five new inlets were formed in the relatively small barrier beach systems of Massachusetts alone. This process may be particularly significant along the extensive barrier beach systems of the East and Gulf coasts.

Like Pleasant Bay, the estuaries protected by these barrier beaches will experience sea level rise as a change in tidal range, rather than as a straightforward rise in sea level. The increase in tidal range will increase the flushing process and that will improve water quality, stimulate marsh growth and the overall productivity of the estuaries. If this happens, we may just see the creation of new lagoons and more "clam factories" up and down the world's coasts.

CHAPTER 25

"The Ink's Still Wet"

November 1992

Su-Ann Armstrong had a cold. It was November 5th, the day after the election. She was sitting at home watching C-Span, searching for any clue that the defeated president might sign a bill critical to the future of the Chatham fleet.

Two weeks before, Mark Forrest had called to tell the Chatham Fishermen's Wives Coalition that Congressman Studds was spending his Sunday huddled in a room with Ted Kennedy, John Kerry and Pat Moynihan. The four Democrats were chortling over their new strategy. They were going to attach language calling for the dredging of Chatham Harbor to President Bush's Fisheries Trade Agreement with Estonia.

"It's a shoo-in," gloated Studds' aide. "He'll have to sign it. It will be his first treaty with a breakaway republic of the Soviet Union, and he authored it."

Su-Ann was leery. The coalition had been disappointed too many times before. The same amendment had been attached to two former bills. Both had died in Congress.

Most importantly, the clock had been ticking fast. Congress had passed the bill just before the election. If Bush didn't sign the bill, it would be pocket vetoed automatically in ten days. The days had gone by quickly and Su-Ann feared that with his defeat looming before him the President would have bigger things on his mind.

During the election Mark Forrest had assured the coalition

that the bill was still secure. "The President doesn't want to be criticized for being preoccupied with foreign affairs. He'll sign it after the election."

But the election came and went. There were only two days left before the pocket veto would go into effect.

Suddenly C-Span broke into it's regular programming. The President and Mrs. Bush were returning from Houston. Their helicopter slowly loomed out of the mist, circled the White House and settled down quietly on the South Lawn. The huge rotor continued to spin as Mr. and Mrs. Bush appeared at the helicopter door. The White House staff cheered as the Bushes hurried toward the side door of the home they had just lost.

Su-Ann searched their faces for clues about how the Bushes were feeling. She had voted for Bill Clinton but could still feel sorry for the President. George Bush looked pinched and tired; Barbara looked relieved. Millie, their dog, was just happy to have her ears scratched by the newly defeated president.

A few more steps and the First Family was gone, safely ensconced inside the White House and away from the cameras.

Su-Ann got up to switch off the television set. It was 3:30. There was no way the President would sign the bill today. In two days it would be dead. Chatham had lost again.

Thirty minutes later the telephone rang. It was Congressman Studds' office. "He signed it. The president just signed the bill. The ink's still wet on it."

Su-Ann blew her nose and rushed to the marine phone. "Chris, we did it. Bush just signed the bill. Tell the rest of the fleet."

Nick Brown remembered feeling happy but also somewhat let down when he heard the news. "Somehow it seemed like we had won by a trick. I would have preferred to win by the numbers but I guess 'All's well that ends well.'"

By the following Sunday, Su-Ann's cold was turning into pneumonia. She was supposed to coach the fishermen's team in the First Ever Annual Sink or Swim Softball Tournament. The temperature had plunged into the thirties, and rain and sleet drenched the VFW baseball diamond. By one o'clock Mark Forrest hadn't arrived to referee and only four of her players had shown up for the game.

"Well, Su-Ann, I guess it's time for us to break out the victory beer." It was Jack Downey, captain of the Coast Guard team.

"Not on your life, Downey. Where's the damn phone?"

Within ten minutes Su-Ann had her entire team dressed in bright yellow oilskins and ready to play. Not to be outdone, Downey suited up his Coasties in their bulky orange neoprene survival suits.

Su-Ann selected Mark Farnham to arm wrestle Downey for first ups. She whispered into Mark's ear, "Squeeze his bloody wedding ring."

The strategy worked. Downey yelped in pain and Farnham slammed his hand to the table.

Everyone agreed it was fitting weather for a Chatham baseball game. Shareen Eldredge drove in the first run and the final score was Fishermen 21, Coasties 11. It was posted right up there on the scoreboard between a Chatham Codfish and Shareen's handmade flag of Estonia.

But the fleet still harbored doubts about the dredging because the Army Corps of Engineers had fought them so hard before. Would they really just roll over and play dead because of this new bill, which had not actually appropriated any money to do the dredging?

Two weeks later their doubts were dispelled when the dredging committee met with the Army Corps of Engineers. There were claps on the back, handshakes and smiles all around. All that was left was the paperwork.

A few days later a letter to the editor appeared in the *Chatham Chronicle*. It pointed out that the Estonian Fisheries bill highlighted the need for a presidential line item veto to avoid pork barrel politics and keep down the deficit. The argument was not well received on the Chatham docks.

CHAPTER 26

"Life Ain't Fair"

—Governor William Weld,
April 1993

It was a warm day in winter, one of those rare and cherished events on Cape Cod. The sun had discovered a break in the relentless days of gray and cloudy skies. It shined down on the quiet waters of Pleasant Bay, rippled only by the splashings of a pair of swimming buffleheads. It was a good day to reflect on the past year and the changes wrought by the inlet of '87.

All told, 1993 was a quiet year. True, the winter was several degrees cooler than normal and the summer was warm and dry—one of the best in memory. There had been persistent heavy snow storms in February and March, but snow on Cape Cod provides a much-needed respite from long spells of rain and cold.

A February storm dropped twenty-four inches on the Outer Cape. The next day the sun sparkled on the new-fallen snow so seemingly incongruous beside the cold green waters of the Atlantic. Cross-country skiers traversed the high dunes of Truro and Ballston Beach, while surf lapped at the sands below.

The Blizzard of 1993 struck in March, close to the anniversary of the Ash Wednesday storm that so devastated the East Coast in 1929. The '93 blizzard dumped a record amount of snow in a broad swath from Alabama to New Brunswick. But the brunt of the storm hit other areas. Erosion was not severe on Cape Cod.

In fact, the dominant change in Chatham was a buildup of sand in 1993. When Bert Nelson left for Florida in autumn, the Atlantic Ocean was 19 feet from his house and the new revetment just south of the Chatham Light. The following spring the beach was 300 feet wider, as thick as it had been prior to the formation of the inlet. To the north, a broad new beach glistened in front of the new light-house revetment and the tombolo stretched to South Beach.

Large flocks of eider ducks discovered Chatham in mid-winter. Huge white rafts of the bobbling, squabbling birds gulped down bushels of mussels that had reestablished themselves in the new lagoon south of the inlet.

Landings of mussels provided a good barometer of the chang-ing productivity of the area. Prior to the new inlet, landings had been moderate. The first year after the inlet formed there were no landings; all the mussel beds had been smothered by in-washing sand. By 1992, improved conditions had brought the catch up to 100,000 bushels, providing income to support close to 40 fisherfolk. Steamers in the new lagoon continued to support an-other 500 shellfisherpeople in 1993.

Landings of groundfish, such as cod, haddock and flounder, were down by 20 percent at the Chatham Fish Pier. Shoaling through the inlet had convinced captains of the larger boats to move to other ports, but part of the decline was due to the collapse of the fisheries themselves.

The greater Georges Bank area used to be the fourth-largest fishing ground in the world. However, the creation of the 200-mile fishing limit in 1976 attracted more domestic money, boats and fishermen than the stocks of fish could support. By the end of 1993 the stocks were close to a total collapse and hundreds of thousands of families in Canada and the Northeast would soon be on unemployment and facing the loss of boats, homes and their life's savings. It was an unprecedented environmental tragedy that we allowed to happen in our own backyard while we were preoc-cupied with spotted owls, rain forests and global environmental problems in more exotic climes.

In late winter the ribs of an ancient schooner emerged from the sands of North Beach. It was a reminder that we have been living on borrowed time. The dire predictions of boating fatalities that had been made immediately after the inlet opened did not occur right away. There had been some earlier close calls; when Jack Our fell overboard while unloading his skiff in Chatham Harbor, and when a summer swimmer broke his neck diving into the treacherous waters off South Beach.

But in early February of 1993, the Chatham Coast Guard station received a call from a fishing vessel in distress. It was 10:35 P.M. The *True Life* was swamped and wallowing in 15-foot seas 30 miles southeast of Chatham. Several Coasties piled into the 44-footer and headed through the inlet into the teeth of 10-foot waves. A Falcon jet screamed overhead and a steady "whop, whop, whop" gave assurance that a Coast Guard helicopter was also lumbering toward the sinking vessel.

While a Boston-bound freighter provided illumination, the helicopter team lowered two auxiliary pumps. One wouldn't start, the other stalled. The vessel couldn't be saved. The pilot used his loudspeaker to order the three men on deck to put on their survival suits and board the life raft. Thomas Deegan had already climbed into the raft when the *True Life* capsized, throwing all three men overboard.

The pilot lowered a rescue swimmer into the frigid waters. After several attempts, he was able to locate two fishermen and slip them into hoist harnesses. They were flown back to Cape Cod Hospital, where Deegan was declared dead on arrival. Several days later another fishing boat found Eamon Harrison, an Irish national, dead inside the life raft. Evidently he had been hiding below decks during the rescue. Only Calvin Lilliston survived. The 44-footer had played no part in the rescue.

Another freak accident occurred on a beautiful day in early June. Roger Ling was jet skiing in the inlet while his stepfather Tom Waage and Samuel Hopkins flew overhead. Suddenly the small stunt plane stalled, rolled, and then hit the water with a burst of spray. Roger sped to the spot and found Samuel Hopkins flailing toward the surface with blood and foam streaming from his nose and mouth. Ling circled the spot in a desperate attempt to find his stepfather. All he found were shoes and a cap floating on the surface. Thomas Waage had drowned in the fuselage of the wreck as it sank below the surface.

However, the most momentous boat ride of 1993 lasted only fifteen minutes. Chatham officials had finally convinced Governor Weld to come and see some of the problems of the inlet first hand. He arrived in mid-April and was greeted with mid-April weather. His boat was jam-packed with dignitaries and photographers. The unflappable governor pointed out that the trip was rather calm, compared to his recent white water canoe trip.

Of course, the governor had the leeward seat. His new Secretary of Environmental Affairs did not. After 15 minutes of cold, drenching salt water, Trudy Coxe had had enough. "Get me outta here. Give Chatham anything it wants."

After viewing the new revetments, Governor Weld observed that "Life ain't fair. It seems that the difference between a bank and a dune is determining a lot of things." The words of both Weld and Coxe would prove prophetic.

In December Aubrey Consulting Company submitted Phase II of the Environmental Impact Review for Chatham's Little Beach area. It was an extraordinary document that would have been unthinkable only a few years before. It called for a series of seawalls and groins to protect the sixty houses on Little Beach and revetments to protect the causeway that provides access to another sixty houses on Morris and Stage islands.

The document pointed out that South Beach would break up during the next 5 to 30 years, exposing Little Beach to the direct force of the Atlantic Ocean. However, it also showed that sand from South Beach would migrate to the mainland and could be used as a temporary buffer to protect the coast while North Beach was regrowing. Temporary wooden groins would be used to trap and hold the sand from migrating south, while a 10- to 12-foot seawall would help protect the homes.

The document was revolutionary. It was the first time such an EIR had included a cost-benefit analysis that looked at the worth of homes and how much tax revenue they contributed to the town. The sixty houses on Little Beach are worth $23 million dollars and provide $147,000 in annual tax revenue. Over the next decade that will represent $1.4 million dollars in taxes to the town.

Formerly, the state would not allow the value of buildings or land to be taken into account when making decisions about slowing erosion. Town and state environmental officials were only allowed to consider the value of the sand in the lots that would be carried away and be deposited on downstream beaches.

The recommendation to build a seawall behind the existing dunes was clearly a way of getting around the state's dune/bank distinction. If the state approves this recommendation, it will echo Governor Weld's words and be a tacit admission that decisions based on the dune/bank were often capricious, unfair and downright ludicrous.

Finally, the EIR proposed the use of groins. Only a few years ago the recommendation to use groins would have been unthinkable. The reason that groins have been repudiated is that they work too well. While they protect upstream beaches, they cause erosion to downstream beaches by cutting off the flow of sand. The EIR calls for the sequential construction of temporary groins as certain trigger points are reached. They will hold sand in place as South Beach breaks up, but be removed to allow the beach to return to equilibrium as North Beach grows to provide new protection.

Dave Aubrey's document shows how far we, and he, had come in understanding the complexities involved in dealing with coastal erosion. Six years after the '87 inlet we were starting to understand that coastal erosion is not a simple problem requiring simple answers. "Let nature take its course," or "Shore up the entire coast" might make good slogans for bumper stickers but they make flimsy arguments upon which to base policy decisions.

CHAPTER 27

"Look Before You Leap"

December 1993

The day after Christmas I stumbled downstairs, wondering how to jumpstart my engine. It's hard to face an empty computer screen after so long a hiatus. But there on the front page of the *Boston Globe* was the impetus I needed. The headline read, "Global Warming at Center of Libel Suit."

It was too good to be true. Global warming was about to become a good read. The article was nasty, personal and replete with public figures throwing mud balls, albeit intellectual ones.

The case involved the reputations of Vice President Al Gore, conservative columnist George Will and the recently deceased scientist Roger Revelle. At issue was the authorship of a paper entitled, "What To Do About Greenhouse Warming; Look Before You Leap." It was signed by S. Fred Singer and Roger Revelle.

The paper opened with an overview of the difficulties of predicting future temperatures and pointed out that, even if greenhouse warming does occur, the net result may well be beneficial. It also suggested that government should pursue policies such as energy conservation "that make sense even if the greenhouse effect did not exist." That sounded like good old common sense from where I sat on Pleasant Bay.

The problem was that Roger Revelle had been Al Gore's mentor at Harvard College. He inspired the Vice President to write his best-selling book, *Earth in the Balance*. Among other things, the book calls for a multibillion-dollar Global Marshall Plan to halt the greenhouse effect.

George Will had used the Singer-Revelle paper to question Gore's environmental views. He had wondered out loud that, if the Vice President's mentor is no longer worried about global warming, should we be? Should we spend billions of dollars, alter the economic and social lives of five-and-a-half billion people, and risk causing political instability to solve a problem whose cure might be more dangerous than the disease?

Gore had been concerned enough to ask Justin Lancaster, a Lexington oceanographer, to write a rebuttal that charged that Singer had taken advantage of the dying scientist. Lancaster had claimed, "Roger did not help write that paper, nor does it represent his views." Now Dr. Singer was suing Lancaster for libel.

Of course the suit was unfortunate. It had turned a much-needed scientific debate about global warming into a nasty political feud.

Roger Revelle had been a careful man who often pointed out that he was a scientist first, and an environmentalist second. This caution often put him way ahead of the learning curve of his colleagues. He used substantiated numbers to unearth fallacies of environmental theories that happened to be in vogue at the moment.

From my student days in the Sixties, I remember when Dr. Revelle wrote a paper that showed that the problem of world hunger was a problem of distribution not production. I remember being aghast; we had all been taught that population would grow exponentially while food production grew arithmetically. It had to be a problem of production. Of course, Revelle was roundly criticized for his views, but eventually it became clear that he was correct.

Revelle was also a staunch defender of nuclear energy at a time when it was out of vogue with most environmentalists. I found it totally within character that Dr. Revelle would coauthor such a cautionary paper and expected that he might very well be correct again.

The libel suit was particularly interesting because it reveals some of the pitfalls about the way we think and write about science.

Science and politics are different. In politics you can go back and forth, debating about liberal and conservative policies forever. It is understood that the debate will never be resolved; it is simply a difference of opinion and depends on who is in power. Ironically, this makes consistency more important in political and legal debates than in scientific ones.

In science, it is assumed that if we collect enough data and analyze it carefully enough we will eventually get closer to the truth. Global warming provides a good example.

When Roger Revelle convinced his colleagues to start measuring the accumulation of CO_2 in the atmosphere in 1957, it made sense to assume that the world would react to rising CO_2 like a greenhouse. We now have over 30 years of studies that show that the world is not as simple as a greenhouse. We have discovered confounding feedback mechanisms and much uncertainty. The cautionary tone of Singer and Revelle's paper simply reflects that realization.

There is another difference. How long would a lawyer be in business if half-way through a case he turned to the jury and said, "Gee, maybe the defense has a point"? How long would a politician be in business if he got caught flip-flopping too many times? That sort of thing is expected, even encouraged, in science. It is an integral part of the scientific method.

The media, and science writers in particular, too often treat scientists as though they were rock stars. There is too much concern about who wins a Nobel prize and who supports a particular theory. Some of the world's most respected scientists have come up with totally wacky ideas. Their ideas should not be allowed to rest on the reputation of the scientist but on the provability of the idea itself. Even the lowliest graduate student should be able to topple the ideas of an established scientist by producing convincing evidence from an enlightening experiment and a few carefully derived numbers. Science is a gloriously democratic endeavor.

People also become disciples of the pioneers in their field. This happens in modern science as well as in movements created by scientists in the past. The disciples try to protect the reputation of their leaders by disavowing papers that their leaders once wrote that contradict tenants of the field's central dogma that were established later.

Usually the pioneers of a field have a greater respect for the limitations of their analysis than their followers do. Perhaps we could have been spared some of the fallacies of Marxism, Freudian analysis, and social Darwinism if we had listened to some of the doubts of Marx, Freud, or Darwin rather than the dogmatic certainties put forward by their followers.

The environmental movement is now at just such a crossroads and its central tenants are being questioned. Global warming has become a central dogma of the environmental movement, but we are discovering that perhaps the greenhouse effect is not as serious as we once thought.

Only a few years ago scientists were predicting that the sea level would rise 30 feet during the next hundred years. They have quietly lowered those estimates to less than one foot, but many in the environmental movement keep using the old numbers and raising the old alarms.

One of the reasons that people are beginning to doubt physicists' computer models about global warming is that they are not standing up when they are compared with data from the real world. Some of the differences stem from the ways various types of scientists collect and interpret the data. When geologists such as Dave Aubrey go out to actually measure how much the sea level has been rising, they discover that the coast has also been sinking. On Cape Cod, he discovered that half of the one-foot rise in sea level over the century has actually been the result of the coast sinking by six inches as the coast readjusted to the retreat of the glaciers. When the weight of the glaciers was gone, the interior of the continent rose, but the coast at the edge of the continent sank.

Other scientists have suggested that up to a third of sea level rise comes from pumping aquifers and increasing runoff due to deforestation. Some have even suggested that global warming could lead to falling sea levels. They reason that global warming could lead to more snow storms in the Northern hemisphere, which would lock up ocean water in glaciers.

Ecologists have discovered similar complexities. When they have gone out into the field, they have discovered that the world does not react the way physicists' models suggest that it should. Despite a dramatic increase in CO_2, the planet does not seem to be warming as much as expected. Productivity in individual ecosystems such as Pleasant Bay may not decrease but may actually increase because of global warming and sea level rise.

Perhaps a reassessment of the greenhouse effect is in order. If so, we would do well to listen to the doubts of Roger Revelle rather than to claims that he never uttered them. To do otherwise would be to do a great disservice to ourselves as well as to the memory of a thoughtful scientist.

CHAPTER 28

The Unraveling

May 1994

The winter of 1994 is finally over. Ducks still rest on Pleasant
Bay and harbor seals still gorge on incoming herring. But the
snow and ice have almost disappeared. It is the end of another
season; and it is time to take stock, time to see if there are some
conclusions to be drawn from the past seven years since the inlet
broke through.

It is also a time to envy the novelist. He can wrap up the loose
ends and finish the story however and whenever he wants. We
must leave the story of Chatham somewhere in the middle. But
perhaps a pattern has emerged and perhaps a few themes may endure.

On January 26th a steady wind blew out of the Northeast,
dislodging the solid pack ice of Pleasant Bay. The outgoing tide
entrained boat-sized chunks of ice and swept them toward
Chatham Harbor. Multi-ton ice floes bore down on moored fishing boats. Hulls creaked, lines strained and all hell broke loose.

One by one, mooring lines snapped and people's livelihoods
drifted toward the inlet. Two skiffs were flipped and swamped by
the fast-moving ice. The fishing boat *Christen* dragged her six-
thousand-pound mooring. She grounded on a shoal and a passing
ice floe slashed a two-foot hole in the side of her hull.

The Coast Guard used its rescue boats as ice breakers to reach
the rapidly drifting vessels. Two Coasties donned survival suits, ran
over stalled ice cakes and through knee-high puddles to reach the

Bearded Clam just before she was dragged through the inlet by the drifting ice. By the end of the day the Coasties and captains had moved twenty boats to the Chatham Pier to protect them from the crushing ice.

It was yet another day in the winter that would not end. In all, sixteen storms battered the Northeast. America became a nation of New Englanders obsessed with the weather. Dave Letterman devoted an entire program to pondering the difference between a cold snap and a cold spell. The *New Yorker* carried a cartoon of a snowbound New Yorker entering his apartment, stamping his feet and heroically reporting to his wife, "They're burning pianos in Hartford."

Boston's Mayor Thomas Menino griped on television, "I've had more snow during my first two months in office than Mayor Flynn had to endure in ten years." The *Boston Globe* started running a daily front-page picture of the seven-foot-tall Boston Celtic player Robert Parrish and compared the rising snowfall against the basketball star's height.

By the end of winter a snorkel had magically appeared in Robert's mouth to help him breath because the snowfall had reached eight feet. Boston had received more snow than Buffalo and had broken the record that had been set fifty years before.

As spring approached I watched the newspapers with rising anticipation. Would we get the last seven-tenths of an inch of snow necessary to break the record? I didn't dare admit it to my neighbors, but I was secretly hoping for just enough snow to break the record set in 1987—the year the inlet was formed.

I was rewarded four days before the arrival of spring. It was St. Patrick's Day morning and outside beautiful, thick white flakes of heavy snow swirled around our house. One of those blondes on the weather channel—the incessantly perky one from Georgia, I think—verified my excitement. Cape Cod would indeed break the record set at the Chatham Weather Station in 1986–1987.

The record would give this book a certain sense of symmetry—the illusion of a beginning, a middle and an end. But of course it was a false sense of symmetry, because the weather is a rogue. A year of bad weather is but a blip on the radar screen of climate. In terms of long-term trends and cycles, a year of colder, stormier weather means nothing more than providing fodder to confirm one's worst fears or favorite theory.

On March 14th, *Time* magazine carried an article that reported that insurance companies were starting to refuse to write policies on coastal homes because global warming threatened to bankrupt the insurance industry. At the Post Office in Woods Hole, scientists scoffed at such articles as they needled their colleagues who most vociferously espoused global warming.

Dave Aubrey had just returned from Russia, where he had read about the East Coast storms on a Moscow street corner. He pointed out that it was only a slight alteration of the jet stream that had funneled the storms over Washington, New York and Boston—three cities that formed the information corridor of the planet. No wonder the world was thinking like New Englanders. Every morning, East Coast reporters had to dig their cars out of eight more inches of new snow.

Unlike reporters, scientists have to base their forecasts of trends on more than one, two or even three years of bad weather. Dave mentioned research that showed that a system of North Atlantic currents acts like a conveyor belt to remove cold water from the surface of the North Atlantic Ocean and carry it down toward the tropics. The system tends to recalibrate and modify the climate over a seventy-year cycle.

For the first twenty years of this century, water off the East Coast was cool. Then it warmed for about ten years, stabilized for about twenty years, cooled for about twenty years and then stabilized again for another twenty years before repeating the cycle. When the waters off the East Coast are cool, the system tends to shift storms north into Canada; when they are warm, it tends to funnel them up the East Coast of the United States.

We are still in a period of cool waters and can expect these temperatures to rise and affect the course and frequency of storms along the East Coast. However, these are natural variations that are not caused by human activities or global warming. It might be prudent for insurance companies to cut back on insuring coastal homes, but their industry is not going to go belly up because of global warming.

In fact, Dave Aubrey thought that natural variation would eventually lead to a natural cooling of the climate similar to what happened during the Little Ice Age in the 1600s. Bob Oldale, his colleague at the Woods Hole branch of the United States Geological Survey, also thought that the present rate of sea level rise could not be sustained and would gradually taper off. We chortled that humans might be inadvertently doing the right thing by warming the climate to counteract a natural cooling.

George Woodwell of the Woods Hole Research Center was incensed. He thought that the article in *Time* magazine was right on. It was an article in the *Boston Globe* that highlighted the uncertainties about global warming that raised his ire. Civilization as we know it was about to end in the next 50 years and newspapers were writing about uncertainties?

"That's the trouble with scientists," said Andy Solow, an economist at the Woods Hole Oceanographic Institution. "They have blinders on. They think because it's science and because it's in their field it's the most important thing in the world. If you listed the twenty most important problems facing the world today, global warming wouldn't even make the list."

On May 24, Justin Lancaster retracted his allegations that S. Fred Singer had coerced Roger Revelle into writing their cautionary paper about global warming.

What are we seeing here? It is clear that our understanding of global warming is a moving target. Many scientists now think that the string of hot summers that plagued the Midwest in the 1980s were also caused by natural changes in the size of patches of warm water in the Pacific, the El Niño-La Niña effect, not by global warming.

Unlike the ozone hole, which seems to get worse every time

scientists investigate it, every time someone looks at global warming it seems to become more confusing and more uncertain. Are we seeing the unraveling of a beautiful theory by an ugly fact or, in this case, by several ugly facts? If so, it does not mean that the original science was bad, but only that the scientific method is working.

Perhaps the most interesting question is the anthropological one: Why do humans want to blame natural events on human conduct and why do they want to believe that they can control them? When the inlet broke through in Chatham, some people wanted to sink ships in the breach to seal it. When Mount St. Helens erupted, people wanted the Air Force to drop bombs down the throat of the volcano to quell it. To alter global warming, researchers have experimented with dropping tons of iron filings into the Pacific to increase the number of plankton that gobble carbon dioxide. A gardener in France once tried to convince me that "If you Americans would just put those rocks back on the moon, the weather would return to normal."

After the earthquake in Los Angeles, an expert on the Larry King show insisted that the tremor was caused by all the things humans were doing to pollute the environment. Another guest was quick to point out that polluting the environment had nothing to do with earthquakes and the only thing that humans are doing differently is that more of them are living in earthquake-prone areas.

The same is true for the coasts. The amount of erosion occurring on the coasts is no different today than it was a hundred years ago and is probably no different than it will be during the next hundred years. The only difference is that there are many more people living on the coasts today than there were a century ago.

Right now, all that can be said about global warming and sea level rise is that we don't know if, or when, or how human activities will affect our global climate. We don't know whether the climate will warm up or cool down. We don't know whether warming or cooling will raise or lower sea levels and we don't know whether rising or falling sea levels will increase or decrease marshes and estuaries.

However, the best guess is that during the next hundred years the climate will continue to oscillate and recalibrate naturally. The world may warm slightly due to human activities, but the planet's natural feedback mechanisms and cycles will tend to alter and mitigate the changes. The sea level will continue to rise at its present rate of one foot per century. That rise will continue to cause erosion problems for people living on the coasts, but they are not problems we cannot handle.

It makes common sense to discourage people from building in erosion-prone areas and it makes common sense to gradually purchase those homes that are in greatest danger. It makes very little sense to drastically alter the economic, political or legal systems of five-and-a-half billion people to meet the challenges of global warming before we really know what they are. It is time to replace the elegant theory of global warming with a new—perhaps less beautiful—theory that better fits the ugly facts.

The Seventeenth Anniversary of the Inlet

January 2, 2004

It is January 2, 2004, the seventeenth anniversary of the storm that opened the Chatham Inlet. Townspeople cluster below the Chatham Light hoping to catch a glimpse of a fishing boat picking its way through the inlet. The main channel or thalweg of the inlet has somewhat inexplicably shifted north as North Beach itself has continued to grow southward.

However, the greatest area of concern lies out of sight several miles south. There, the tip of South Beach is reaching out like the finger of God to join up with Monomoy. When it does, the event will have repercussions from Anchorage, Alaska, to Washington, D.C.

The reasons are these. According to law and local custom, if you happen to own a piece of land that is eroding, it is simply your loss; conversely, if your land is accreting, it is your good fortune, and you own the new land. So, when South Beach attaches to Monomoy, Chatham could argue, "Okay now the Monomoy Wildlife Refuge is part of Chatham." Conversely, the Refuge could just as easily argue that, "Okay, now South Beach is part of the Monomoy Wildlife Refuge, and thus it comes under the provisions of the Wilderness Protection Act." All this is further complicated by the fact that the Cape Cod National Seashore has just as legitimate a claim to South Beach as either of the other two parties.

But, here is the real kicker. The Wilderness Protection Act is the same piece of legislation that has prevented oil companies from drilling for oil in the Arctic National Wildlife Refuge. The reason is that one of the provisions of the Act prohibits commercial activities in designated wilderness areas. In Alaska this has meant no drilling for oil, but in Chatham it could mean no digging for clams on Monomoy or in the "Clam Factory" that has become a local bonanza bringing in four million dollars a year and providing several dozen families with 95 percent of their income.

The genie jumped out of the bottle in 2002 when a Japanese company sued the federal government to allow it to continue harvesting horseshoe crabs within the Refuge for the lucrative lysate industry. The lawsuit backfired. Scientists not only discovered that harvesting the crabs for lysate decreased the population of this uniquely valuable biomedical resource whose blood is worth over $15,000 a quart, but they also discovered that harvesting crabs also jeopardized the often-endangered shorebirds that feed on horseshoe crab eggs—and the Refuge was created to protect these migratory birds. With shellfish the reverse is true. Scientists have shown that fishing for shellfish actually increases the productivity of clamflats by turning over the sediments, removing older clams and increasing the number of smaller, younger clams, which provide food to most of the shorebirds. One species of shorebird, the handsomely marked American oystercatcher, has become significantly more abundant on Monomoy since shellfishermen started working the "Clam Factory" after the inlet opened.

So the scientific argument is fairly clear-cut, but the legal arguments remain murky. The problem is precedent. The great fear in Chatham had been that some environmental group would get it into its head that if the federal government allowed shellfishing on Monomoy, under the provisions of the Wilderness Act it would also have to allow drilling for oil in the Arctic National Wildlife Refuge!

Legally they had a point. The thrust is the same—commercial activities should not be allowed within a wildlife refuge—but can you really equate a guy with a rake going out to scratch up a few clams with an international corporation going out to install a multi-million-dollar rig to drill for oil?

Up until January 2nd, the forces for common sense seemed to be in control. Both sides wanted to avoid the litigation and acrimony that had been engendered by the horseshoe crab decision. The town of Chatham quietly paid a Washington lawyer a quarter of a million dollars to provide legal and scientific arguments to the U.S. Fish and Wildlife Service to avoid a legal showdown over shellfishing on Monomoy. For its part, the Fish and Wildlife Service also seemed to want to avoid litigation and appeared to be looking for ways to make a ruling that would continue to find shellfishing compatible with the purpose of the Refuge, which they own and operate. Each side was hoping that their hands would not be forced by outside pressures or, worse yet, be caught up in what appeared to be becoming a bruising presidential debate about the future of the Arctic National Wildlife Refuge. The inlet had shown once again that nature still had the ability to reach down and shake up the human world.

But on January 2nd, Chatham heard the news it was most hoping not to hear. An article in the *Anchorage Daily News* announced that the 9th Court of Appeals had ruled that salmon could no longer be stocked in the Arctic National Wildlife Refuge because stocking was considered a commercial activity. This was not some environmental group; this was the U.S. Court of Appeals.

The Alaska decision was certainly relevant to the case that had banned collecting and returning horseshoe crabs to Monomoy, but it also had an uncomfortable relevance to the benign, even beneficial, practice of shellfishing on Monomoy. Both Chatham and the U.S. Fish and Wildlife put on a brave face about the ruling, indicating that the Monomoy decision would be based on different science and that nothing would happen for at least two years. But it was clear that the federal government's hand had been moved by a decision made thousands of miles away in Anchorage.

Only time will tell if Chatham can save its traditional shellfish industry that had been vastly improved by the creation of the inlet in 1987. However, the real threat continues to be, what will happen to Chatham when South Beach no longer offers protection against the full force of the Atlantic Ocean? Many people are surprised that this has not already happened. Remember Chatham's 140-year inlet cycle? When the previous cycle started with an inlet opening in 1846, the worst erosion occurred after South Beach broke up—some forty years later. But this time the breakup of South Beach could affect up to a hundred homes. As of January 2, 2004, we are halfway to that point of worst erosion. When South Beach breaks up, will we be any better prepared than we were in 1987?

There is some reason for hope. In 1992, the Supreme Court had decided in *Lucas v South Carolina Coastal Commission* that a state had to compensate a homeowner if the state's environmental regulations lead to a reduction in the value of the homeowner's property. In 1994, the Supreme Court had decided in *Dolan v Tigard, Oregon* that the state had to prove the scientific validity of its regulations. They had, in essence, turned the issue on its head, shifting the burden of proof from the homeowner to the state. After much deliberation, the Supreme Court had finally decided not to hear Nick Soutter's Nelson-Hicks case. But it was clear that Chatham's inlet had made in impact on the national debate about sea level rise. When Hurricane Isabel battered the East Coast and opened a new inlet in the Carolinas in August of 2003, local communities from Long Island to Florida were quick to offer guidelines and relief for repairing the damage and incentives for those who wanted to retreat.

It is clear that our climate will continue to warm, our seas will continue to rise, and we will continue, most likely, to guzzle fossil fuels. It will be up to us to filter the effects of sea level rise through the ever-changing maze of laws, rights, and regulations that govern our liberal democracy. It will continue to be a difficult but necessary evolution.

Index